THE MIDLANDS' METAL

Nickel in the Service of Man

Corrigendum

Page 96
delete quartz
insert galena

Front cover: A Meteor, the first successful jet aircraft, taking off.
Nickel alloys are needed to get jet planes off the ground.

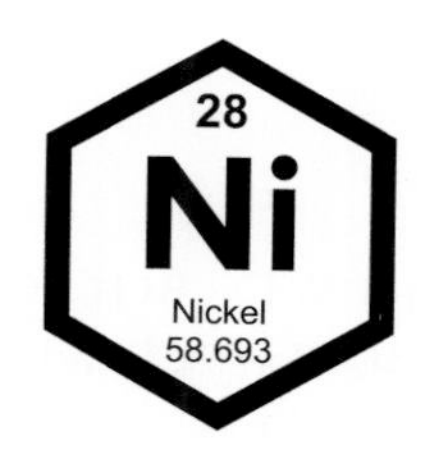

THE MIDLANDS' METAL

Nickel in the Service of Man

Ian Dillamore

BREWIN BOOKS

BREWIN BOOKS
56 Alcester Road,
Studley,
Warwickshire,
B80 7LG
www.brewinbooks.com

Published by Brewin Books 2019

A CIP catalogue record for this book is available from the British Library.

ISBN: 978-1-85858-593-2

Printed and bound in Great Britain by Hobbs the Printers Ltd.

Contents

Acknowledgements

Much advice and help has been necessary in covering the wide range of material behaviour treated in this book. Inevitably the treatment is not as uniform as would be desirable. In particular the section on catalysis is a mere taster to a highly complex subject. However, it was considered important to indicate, as in all other applications, the initial insight that led to the current state of a very important subject.

I am indebted to several former colleagues who are more expert than myself in the subjects covered. Bob Giles substantially revised the chapter on electrochemistry, to which both Peter Crouch and Tony Hart also contributed.

Mike White introduced me to some important new information on stress corrosion and Rex Harris helped on the difficult subject of magnetism. Brian Salisbury made a number of helpful suggestions. Dr Robert Whitworth was helpful in obtaining information on, and illustrations of, magnetrons. David Dulieu not only contributed significantly on the broad issues relating to steels of all kinds but commented extensively on an early draft. Stan Glover and Ray Bishop were similarly helpful on matters of style and substance. I am also fortunate that my wife consistently improves my grammar and tolerates the time I spend on the fun part, which is the research.

Sources of illustrations are acknowledged in their captions.

Figures

A.8 A white cast iron showing cementite surrounded by ferrite hardened by secondary cementite precipitation.
A.9 Illustrating the interface between the nickel rich matrix and the γ' hardening phase.

Foreword

It was in the nineteenth century that scientific discovery started to reveal most of the material behaviours that affect modern life.[1] This continued into the twentieth century. In the same period materials, other than those that had been known since antiquity, became available in quantities sufficient to permit their uses to be explored.

In the following is a review of the scientific history of this period, describing the key discoveries and crediting their originators. The main focus is on a key ingredient on which many, if not most, of the applications that evolved came to depend.

That ingredient is nickel. This has its own story, which started in earnest in the Midlands of England, and continued to develop there for over a century. As with much of the formative activity that made Birmingham and the Black Country the workshop of the world, the story is now history, but it and the people who made it all happen should not be forgotten.

Although nickel has been identified as being present in ancient artefacts it did not become available in pure form and in quantity until the first half of the nineteenth century. That this happened in Birmingham made it the centre of the nickel industry for many years, and in Europe for over a hundred years.

Many of the subsequent developments in its use were made by researchers in Birmingham and many more by their colleagues in sister organisations in the USA. The material behaviours that these developments enabled, and their applications are described in the following chapters. They cover so wide an area that by concentrating on the one metal, more applications are described than in the widely read book by Alexander and Street: 'Metals in the Service of Man'.

The history of science has recently started to be approached by historiographers. They try to analyse history in terms of the social environment in which developments occurred. A particularly influential book by Jan Mokyr, 'The Gifts of Athena', attributes progress to the interaction between 'thinkers' and 'doers', whom he calls savants and fabricants.[2] A more common view is that progress is the result of the contributions of talented individuals, and it is certainly true that some of those celebrated, Michael Faraday, James Watt, Josiah Wedgwood and many more, could not be put into either the savant or fabricant category, but belong in both categories. In most cases progress is made by those with a fertile brain connected to capable hands.

In describing the developments that are relevant here, it will be evident whether the developments were incremental, which we might say were inevitable and, in their time, or if the insight of an individual was responsible. Both circumstances are equally likely.

The brass industry was, by the time our story begins, mainly based in Birmingham and the Black Country. It was the existing enterprise there that had caused it to come to Birmingham and it was the range of expertise that was then produced that made it the logical place for several metals industries to be located. One evolved into ICI Metals Division, subsequently IMI. Another is the starting point for this account, initially as Evans and Askin, which became the principal source of nickel, its salts and its alloys for many years.

The accounts of material behaviour will not always satisfy the expert and in some cases the inexpert may find the going a bit tough. There are supplementary accounts of material behaviour in the Appendix, intended for those who have not studied physics or metallurgy, but these can be left unread for those who are only interested in the many applications that would not have been developed, or now be available if there was no nickel.

A Baskerville font is used throughout. It was in Birmingham that John Baskerville made his major contribution to the development of printing, first using his improved font in 1757 to print a volume of Virgil.

1
Introduction

The Bronze Age followed by the Iron Age are epochs in the progress of civilisation. They were preceded by a period in which wood, stone, and other naturally occurring materials were used. Ceramics were the characteristic materials that revealed the progress of civilisation. The awareness of the materials in use was always important, but their developments over centuries were with a knowledge, but no understanding of the reasons, for material behaviour. At best, some of the activities of alchemists investigated interactions that did not occur naturally, but it was not until the eighteenth century that some pattern started to emerge that led to a systematic approach to chemical processes. Before then only those metals that occur naturally were known. Even zinc, a common component of the copper-based materials, was not isolated (or at least recognised) until 1746. The metals cobalt, nickel, chromium and manganese were all first isolated in the eighteenth century and most of the non-native precious metals not until early in the nineteenth century. (Rhenium was the last of the stable metals to be isolated, in 1925, having been long sought to fill the last gap in the periodic table.)

The earliest use of these newly isolated ingredients was to mix them with the two common bases, iron and copper. For iron, this needed the high temperature melting furnace developed by Benjamin Huntsman. It was first applied to making commercial alloys by Robert Mushet (1811–1891), who added tungsten, another eighteenth century discovery, to an iron-carbon alloy to produce the first alloyed tool steel. Alloy development continued throughout the nineteenth and into the twentieth century using methods that were familiar from the alchemical tradition. Suck it and see was the investigative method.

The ability to make reasonably well characterised mixtures enabled the systematic variation of physical properties to be determined, but it was not until William Bragg (1862–1942) and his son (also William Bragg) (1890–1971) demonstrated, in 1912, the ability to study the internal structure by X-ray diffraction that scientific research in metallurgy took off.

It soon became common for companies to establish activities that aimed to find new properties and combinations that would generate commercial benefit. Where the industry was highly fragmented, companies got together to share the cost of research. The British Non-Ferrous Metals Research Association (BNF) was founded in 1920, the British Cast Iron Research Association (BCIRA) in 1921 and, after the 1939–1945 war, the activities that had been carried out under the Ministry of Supply to support the war effort, were continued by the Government part funding these industry activities and establishing, in 1944, the British Iron and Steel Research Association (BISRA).

The big steel companies, Brown Firth (The Firth Brown Laboratory), United Steel (Swinden Laboratories) and Tube Investments (Hinxton Hall) all had research laboratories equipped with the means of simulating industrial processes and evaluating the results of their experiments. The new nuclear industry created major research centres with the behaviour of metals under irradiation being one of the main areas of study.

The formal organisation of research in England had first been established at The Royal Institution in 1799. There Davy and Faraday were responsible for many of the early triumphs of the scientific method; they were some of the first people whose gainful employment was in carrying out research. In 1896 the building next door to the existing Royal Institution building in Albemarle Street in London was acquired by Ludwig Mond and given to The Royal Institution to be used as the Davy-Faraday Research Laboratory. Mond was dedicated to the notion that wealth comes from innovation and believed in the innovation flowing from fundamental scientific research.

Mond's legacy was in his leadership of science-based industries, which was carried on by his sons. He and they were responsible for creating two major businesses that have had a great influence on the world in which we now live. The Brunner-Mond chemical business evolved into ICI, while the Mond Nickel Company eventually merged with International Nickel.

The research activity initiated by the nickel company was different from those supported by the other metal companies. It is very rare, probably unique, for a company to establish a research laboratory to find new ways to use a single element, not to exploit any new applications for itself but to make the new possibilities available to the whole world.

Later companies that promoted the use of alloying elements, principally targeting the steel industry, generally collaborated with University departments to research the applications for their materials. Climax Molybdenum founded in 1918 had a close association with the Colorado School of Mines. The producers of the micro-alloying elements, vanadium and niobium that became important in high strength steels after about 1960, also chose to sponsor research in Universities.

It was not the way of the steel companies to make their research widely available. They were each trying to steal a march on the others to gain a bigger market share. The logic for the Mond Nickel Research Laboratory was to increase the demand for its existing nickel supplies. When it embarked on this approach it did not have a monopoly of supply. Any benefits it produced would be shared with International Nickel in the USA and Le Nickel in France.

The Mond Nickel Laboratory was located on Birmingham Heath. This was also the area where research in perfecting the Mond refining process had been carried out, because the raw materials that were required were already available there. Thus, the development of this rare facility was a Birmingham phenomenon. The following chapters will give an account of how the subsequent discoveries of the uses of nickel have contributed so much to our lives. Not all were due to work in Birmingham, but many were and following the merger of Mond Nickel with International Nickel, the credit for a large proportion of the discoveries and inventions were shared between laboratories on both sides of the Atlantic.

Following the merger, International Nickel did become dominant in the supply of nickel and its directors could see the sense in continuing to seek new ways to use their product. They even took the decision to invest in rolling mills to make semi-finished products available, because the carbon steel and stainless-steel makers were reluctant to produce the materials that had higher nickel contents than they were used to, but that was never what they saw as their core business.

International Nickel, now called INCO, continued product research until 1982. In 1981 the company, which had previously had a licence to print money, registered its first loss for many years and realised that because of new suppliers in South Africa and Australia, other producers were now gaining the greatest benefit from their research. The directors decided to terminate product research, but to maintain a process research laboratory near to its headquarters, which was now in Toronto. Subsequently, INCO also withdrew from its downstream businesses to concentrate on its principal source of income, its mines.

The metallurgical research laboratories listed above are now all gone. There is one significant centre in existence in England as part of the University of Sheffield, called Castings Technology International. It employs 60 workers and has inherited the core interests of the British Cast Iron Research Association and the Steel Castings Research and Trade Association (founded in 1953). The existence of this facility reflects the growing importance of cast metals, which will be seen later. Other Universities also continue to carry out metallurgical research as relevant to the training of scientists and engineers, but the heyday of metallurgical research has long gone.

The buildings on Wiggin Street that housed the Mond Nickel Laboratory have been demolished; the laboratory where Carl Langer perfected the Mond-Nickel process was demolished a long time ago. The variety of ways in which nickel affects us are described in the following chapters; many result from discoveries made in Wiggin Street and by colleagues of those who worked there. This book is dedicated to their memory.

2

History on the Heath

Charles Askin *Brooke Evans*

Henry Wiggin *Ludwig Mond*

Figure 2.1a *The principal players.*

George Elkington *Alexander Parkes*

Figure 2.1b *Supporting actors.*

Sheffield steel and Birmingham brass, they say, but it would be more accurate to choose a different alliteration; Sheffield steel and Bristol brass, for it was at Bristol that the methods used in the production of brass were most established in Britain. This was because shipbuilding was the major trade in Bristol, and the most significant use of brass and other alloys of copper was, until the second half of the eighteenth century, in the metal fittings used on wooden ships. The centre of production moved to Birmingham because of the growth of its use in plated articles and toys – any small metal items – following on from developments in work organisation and, therefore, productivity by John Taylor and Matthew Boulton.[1] There was no significant improvement in the methods or processes, the essentials of which were known to the ancients. But in Birmingham and the surrounding parishes the presence of this activity stimulated a great deal of innovation that led, early in the nineteenth century, to it becoming the home of its own metal: nickel.

Although in use since the earliest times in natural alloys of various compositions, until it could be reliably isolated and produced in quantity, nickel made little contribution to technical or social progress. Since the developments in its production, which occurred initially on Birmingham Heath, it would be difficult to imagine life without it. The range of applications that now depend on this metal will be described in the following pages.

King Arthur's sword, Excalibur, had it existed, would have been made of an alloy of iron and nickel. Pieces of metal containing moderate to high levels of nickel with the remainder largely iron have been found occasionally and are from meteorites. One of the largest came to earth in Brazil: the Santa Catarina meteorite. It is estimated that it may have weighed 25,000kg and consisted of about 30% nickel and 70% iron. Such pieces of metal would have attracted interest by often being shiny and not brittle like stone or the more common rocky meteorites. It would have been found that the material could be beaten and shaped. Swords were made from such metal and had much superior performance to common swords made from iron by cutlers, imbuing them with the mystical significance that gave rise to legends like the Arthurian Excalibur.

The origin of metallic meteorites is the same as that of the earth. The earth's core is also a mixture of mainly iron and nickel, elements that have been formed by nuclear fusion from precursor elements; hydrogen, helium, oxygen and carbon under high pressures and extremely high temperatures in massive stars that exploded,[2] throwing fragments across space. Some large fragments coalesced to form planets and smaller stars, leaving the smaller fragments that constitute meteorites and a lot of dust that is encountered from time to time, manifested on Earth by the Pleiades and the Perseid shooting stars.

Most of the nickel that is available is from the earth's core. Although in the composition of the earth it is estimated to constitute about 5.2% of the whole (iron is 85%)[3], in the accessible crust it constitutes only about 80 parts per million.[4] Iron is about 4.1% of the earth's crust, illustrating the fact that iron is significantly more reactive than nickel, the crust being largely composed of minerals like silica, alumina and calcia; oxides of highly reactive metals that make up hardly any of the core.

The nickel that is mined now is from two sorts of ore. Oxide materials, called laterites are found in tropical volcanic regions on the earth's surface. Long after the volcanic activity, they have become weathered and enriched in nickel content. These are the commonest ores. The second source is obtained by the deep rock mining of sulphur containing minerals that are obviously closer to the earth's core. The most significant area from which the sulphide ores are mined is around Sudbury in Ontario, where a nickel-rich distribution forms an irregular basin shape, thought to have been created when a large meteorite came to earth and created a splash zone, throwing up iron-nickel material from the core. Evidently it was not a metallic meteorite but the more common rocky variety, because the nickel concentration is around the periphery of the zone.

Nickel was not identified as an element until 1751, when Cronstedt (1722–1765) separated it from a mineral (since identified as nickel arsenide) that he had obtained from a cobalt mine. Cobalt had been identified in 1739 by George Brandt (1694–1768), but minerals containing cobalt had long been mined for the cobalt blue used in pottery manufacture.[3] Cronstedt also obtained nickel from a material known to Saxon copper-workers as Kupfernickel. As has often been explained, both nickel and cobalt are derived from German words for little devils and the Kupfernickel was an ore that looked as if it should yield copper but produced a useless metal. However, it later became apparent that the problem was not the nickel or cobalt, but the arsenic impurities. There is nothing devilish about either metal, both of which are now essential ingredients in the most demanding of applications.

Before the end of the eighteenth century, any use of nickel was adventitious. There are several examples from antiquity of artefacts that owed their properties to nickel, including the meteoric-iron swords, coinage and a material known in China as Pai-t'ung (which translates as white copper). However, all the artefacts were made of natural alloys; that is the composition depended on what the ore contained. That was copper, nickel and zinc, together with the undesirable impurity, arsenic. Pai-t'ung was highly prized by the Chinese and was known to have been produced around 200BC, which was when smelting with charcoal had become widespread and had initiated the iron-age. The metal generally looks like silver but is even more resistant to tarnishing than silver and would have been considered especially attractive at a time when most metal artefacts were yellow.

The identification of nickel by Cronstedt and its subsequent refining by Bergman (1735–1784) may have been the catalyst for something like Pai-t'ung to be produced in Germany and used for making cutlery, but it was still subject to poor quality. Charles Askin (Figure 2.1) discovered this when he visited Germany in 1824. He purchased some items made from an alloy, which they called Argentan. They proved to be brittle. It was this experience that led Askin to try to develop something better.

Argentan had been developed by Dr. Geitner (1783–1852) in Schneeberg in Saxony at about that time; his process did not produce pure nickel but eliminated most of the arsenic that had been present in the ores. Geitner's method was introduced into England by Percival Johnson (1782–1866) in about 1830. He became the first Englishman to produce what became known as German-silver. The understanding of chemistry advanced significantly in the eighteenth century. The action of processes that had been regarded as magical by the alchemists were now being given rational interpretations, engaging men like Charles Askin, and Percival Johnson in using them. Johnson, the founder of Johnson Matthey, developed a business in assaying jewellery, and became the leading chemist in the processing of precious metals. This remains the Johnson Matthey business interest. Around 1830, when Johnson was producing German-silver the material would have seemed relevant to this activity, but it is not recorded that he continued it for very long.

Howard-White in his excellent book 'Nickel – an historical review'[3] gives an extended account of the various early players in the manufacture and supply of German silver, including details of the processes used in refining the alloy. However, it was Charles Askin who completed the work and it was the company, of which he was a founder, that had the greatest early impact on the subsequent development of nickel and its uses.

Johnson made his German silver from an impure nickel oxide, sheet copper and zinc, giving 55% copper, 18% nickel and 27% zinc. Askin's first product was somewhat similar, and he went into a short-lived partnership with two brothers, Henry and Theophilus Merry, to make the alloy.

This was sometime between 1830 and 1835. In 1830, the Merry brothers were listed in the trade directory as brass founders, and makers of military ornaments and picture frames. In 1835 they had added German silver to their business. However, by 1839 Askin was shown to be in partnership with Brooke Evans (1797–1862) (Figure 2.1) and it was the company of Evans and Askin that became the biggest supplier of German silver.[5/6]

The accounts of the history given by Howard-White, and in the centenary volume of the company give no explanation of the reasons for the termination of the association of Askin with the Merrys, but it almost certainly resulted from an encounter that Askin had with Edward White Benson (1802–1843).[7] He was the manager of the British White Lead Company, located on Birmingham Heath (Figure 2.2), as well as being manager of the British Alkali Company. Both these companies had been started by William Gossage (1799–1877).

Figure 2.2 *The British White Lead Company, c.1830. (Courtesy Bob Greathead)*

The White Lead company was located close to Icknield Port Road; Evans and Askins' manufacturing site was probably close by at Icknield Square at that time.

The British White Lead Company was located beside the canal built by James Brindley, which was completed in 1769. The implication of the drawing in Figure 2.2 is that the ground on the left of the buildings was unoccupied, which is confirmed by a contemporary map. It was this area that Evans and Askin subsequently occupied and the street on the other side of the buildings subsequently became Wiggin Street.

The problem that Askin and Benson discussed was how to finally purify the nickel. What was then possible to produce, by a sequence of oxidation to get rid of sulphur and arsenic and roasting with charcoal to reduce the oxide to metal, was an alloy that contained copper, nickel and cobalt. This could be used to alloy with zinc to produce Argentan, but at that time the raw material was the crude speiss that was used to produce the blue pigments for pottery and it was the objective to separate the nickel and cobalt so that a cobalt-free German-silver could be produced and the cobalt retained for its primary market.

To further separate the elements, the metal was dissolved in hot acid. It was known that the copper could then be precipitated by adding sodium hyposulphite ($Na_2S_2O_3$)[8] but the close similarity of the chemistry of nickel and cobalt left their separation the one challenge to be overcome. Between them Askin and Benson decided that it was worth trying the addition of bleaching powder (calcium hypochlorite – $CaOCl_2$) to the acid solution. They calculated the necessary amount of the powder to add and each subsequently carried out experiments to try out the idea.

One of them had the correct amount that had been calculated; the other found that he had only half as much as they thought was needed; accounts differ as to which of them was short of the material, but since Benson was in a business associated with making soap, it is reasonable to assume that it was Askin who added only half of the calculated amount. It was found that adding the full amount precipitated both cobalt and nickel, but half did the trick, precipitating only the cobalt.

The 1839 trade directory still has the Merry brothers trading in nickel-silver, but Evans and Askin are also shown to be in business. Since Benson died in 1843 (presumably of lead poisoning), the events described were most probably between 1835 and 1839. With this ability to purify the nickel and cobalt, it is likely that Askin wanted to maximise the benefit to himself, rather than share it with the already established Merrys, but he needed finance, which is where Evans probably enters the picture.

The biographies of Evans generally claim that the collaboration with Askin started before Askin joined the Merry brothers. This seems unlikely. It is more probable that he was known to Askin and had made money from his earlier business ventures that he could invest in this new activity.

The ability to separate out the cobalt gave Evans and Askin an additional trade in supplying salts; notably nickel oxide and cobalt oxide. The latter could be used to make pigments for pottery and glass – a trade that Evans knew; one of his businesses was making glass and lead, the materials for making windows at that time. The nickel oxide is readily reduced by carbon and could be sold to alloy producers. Cobalt is half as abundant as nickel in the earth's crust, so it was doubly important to separate it.

The purer nickel that was produced was free from the brittleness that had dogged the early German efforts; the metal could now be rolled, forged and cast. The market for the silver-coloured, tarnish-free metal for use in cutlery grew rapidly and was enhanced further when electroplating was developed, and it was found that German silver was the best base for plating with silver. Evans and Askin came to dominate the market for the metal.

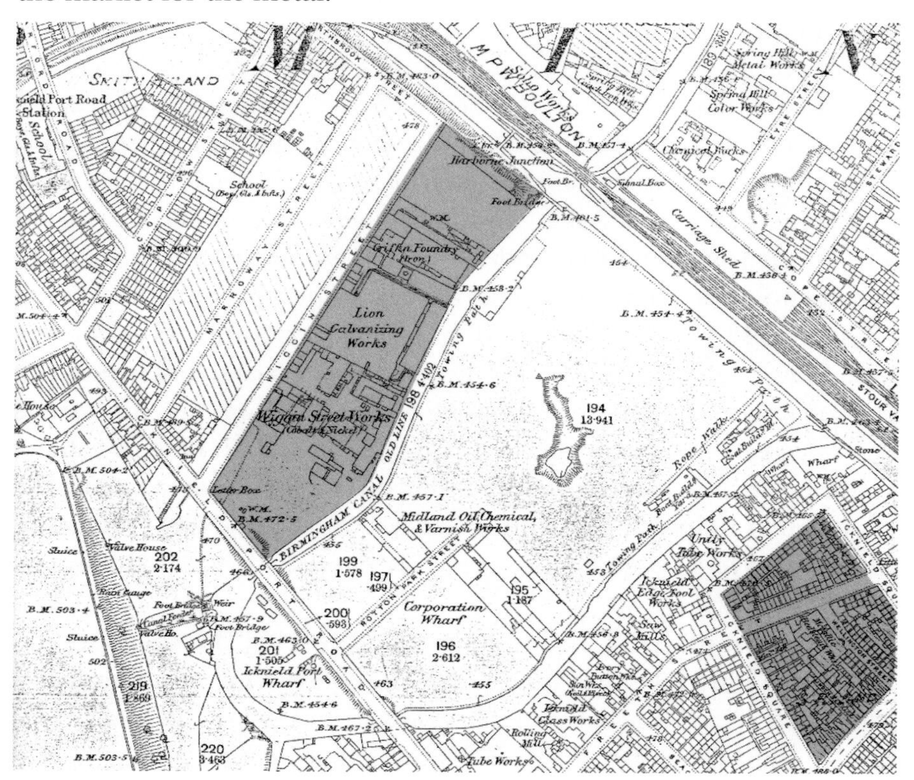

Figure 2.3 *Locations of the business of Evans and Askin, and its successors.*

The company of Evans and Askin were initially located in George Street in the parish of St. Pauls. Their next address was not listed until 1849, when an address on Icknield Square was recorded. This is in the bottom right hand corner of the map shown in Figure 2.3, which is from the 1887 edition of the Ordnance Survey. It is probable that manufacturing moved to Icknield Square before 1849, but with the George Street address being where they had their office.

A further site was added sometime around 1844. The volume issued to celebrate the centenary of Henry Wiggin and Company records that Askin took a young man from his home town of Cheadle in Staffordshire into the business, the eponymous Henry Wiggin (1824–1905) (Figure 2.1), who showed great promise.[9] He was given the task of dealing with the residues from a business that a German had set up next door to the British White Lead Company. Evans and Askin acquired this site in around 1842–1844. The German had been trying to refine nickel and cobalt but without success. Henry Wiggin is said to have extracted enough nickel and cobalt from the residues left on the site to pay the price for the property. The Icknield Port Road site, later Wiggin Street, did not feature in the trade directories until 1862, but became the premises of the company for a hundred years.

To service it they acquired mines in Espedalen in Norway, but none of the European sources were very abundant and it was while Askin was seeking further prospects on a trip to Germany in 1847 that he died. Evans continued his involvement with the company until his death in 1862.

It was, undoubtedly, Evans and Askin's ability to purify the metal and gain the benefit of the cobalt that was also produced that made Birmingham the centre of the industry. Others were attracted to the area to join the trade.[5/8] Stephen Barker set up as a nickel refiner and German-silver manufacturer, as is first recorded in the directory for 1849.

Henry Wiggin eventually took over the leadership of the company, and after the last remaining brother of Brooke Evans died, he changed its name to his own.

The concentration of the nickel industry in Birmingham was strengthened by the emerging use of electrochemistry. This started when George Elkington (1801–1865) (Figure 2.1) patented rights to a process for electroplating silver that he had bought from a Dr John Wright. He started a business that eventually replaced Sheffield Plate in the manufacture of tableware and ornaments. Askin had first encountered German silver used as forks and spoons and with the discovery that the metal was the best substrate on which to plate silver, the demand was boosted significantly. The real genius behind the development of the silver electroplating industry, enabling it to displace Sheffield Plate from the market, was Alexander Parkes (1813–1890) (Figure 2.1).

Parkes was one of the most inventive Brummies of all time. His incredible list of achievements includes: inventing the first synthetic material, a form of plastic he called Parkesine, which was a precursor to Celluloid and Bakelite; enabling the rubber tyre industry by developing vulcanising, and also the raincoat business of Charles Macintosh (1766–1843) by applying the vulcanised rubber to the waterproofing of cloth; developing phosphor-bronze, and, in our context, patenting the first nickel-steel. The nickel-steel was patented in 1870.

The American, French and the Sheffield steel industries soon picked up on the effect of nickel on the properties of steel and the demand burgeoned as the nineteenth century ended. This growing demand for nickel did not go unnoticed by one of the greatest of technological entrepreneurs and it was the availability of the necessary raw materials that brought the next major development in the nickel industry to Birmingham Heath.

In 1889 Ludwig Mond (1839–1909) (Figure 2.1) and Carl Langer, Mond's assistant, had recognised that carbon monoxide was corroding the nickel-plated valves in plant they were constructing to eliminate the gas from streams containing hydrogen.[10] The CO was forming a gaseous compound with the nickel, nickel carbonyl, making available a new option for obtaining a pure form of the metal. In 1892 Mond leased a property

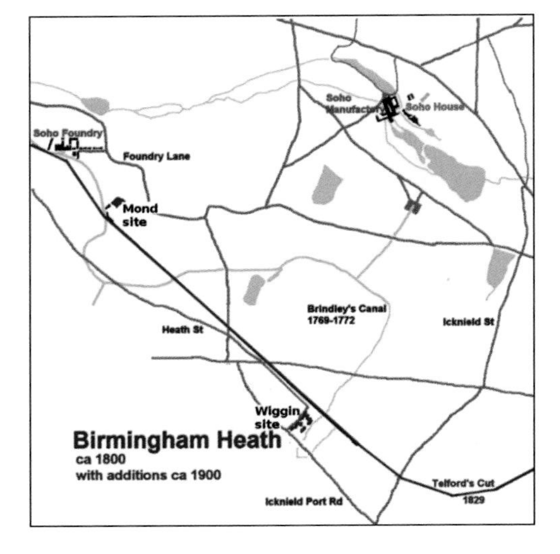

Figure 2.4 *Sketch map of the relationship between the Wiggin sites at the end of the 19th century.*

from Henry Wiggin, located a short distance away from the main plant on Wiggin Street.[11] The location is shown in Figure 2.4. The site where the Mond process was developed is close to the famous Soho Foundry and alongside the Birmingham canal, which connects it to Wiggin Street.

The site was previously used as a soap works and part of it was acquired by Wiggin in 1888. It was used by Wiggin mainly for roasting the ores and boasted one of the three tallest chimneys in the land to disperse the arsenical and sulphurous vapours that this generated.[12]

A small building was erected there to develop the Mond process; a second building served as a laboratory (Figure 2.5).

Figure 2.5 *The buildings where Carl Langer developed the nickel carbonyl process. (Courtesy Peter Crouch)*

The development continued for six years between 1892 and 1898. Mond tried first to interest others in licensing the process before, having failed in this, he decided to operate it on an industrial scale, proceeding to set up the Mond Nickel Company in 1901. The process is described in detail in a paper by Roberts-Austen.[13] It depends on the ability of carbon monoxide gas to pick up atoms of nickel at room temperature and then release them at a temperature of 180-200°C. The constant circulation of the CO is first over finely divided oxide matte containing nickel and copper where the nickel is picked up, but the copper is left behind. The gas is then steam heated and passes over a bath in which seeds of the purified nickel are kept in constant oscillation and the carbonyl vapour deposits the nickel to grow the seeds until at a sufficient size the carbonyl pellets are discharged and replaced with new seeds.

This is essentially the process that is still operated by INCO at its refinery at Clydach in South Wales. This was the location chosen by Mond for its proximity to the steam coal needed to operate the process and to the ports through which the nickel containing matte was imported from the mines he had established in the Sudbury Basin.

These developments changed the UK nickel industry completely. Evans and Askin and then Wiggin had been scratching about in Europe to find ores sufficient to meet the growing demand for nickel and its alloys, but Mond and the Orford Copper and Nickel Company had separately started to extract ore from the much more fruitful Sudbury Basin, while in the same era the French company, which became Le Nickel, was exploiting other productive ores in New Caledonia. (c.1890 Le Nickel had a refinery at Holly Lane, Erdington[14] in Birmingham confirming that Birmingham was at that time a central location for nickel production and demand).

So it was that, having enabled Mond to get started using the materials it had available, and letting it operate on one of its sites, Henry Wiggin and Co had helped to create a powerful competitor for its business in nickel salts for plating, as well as nickel and its oxide for sale to melters, casters and wrought product manufacturers.

In the same period that Mond was exploiting the ability of carbon monoxide to separate nickel from copper, the Orford Copper Company, one of the original constituents of International Nickel, was developing a different approach, using electrolysis. Robert Means Thompson (1849–1930) the first head of the company was responsible for encouraging this development. Victor Hybinette (1867–1937) was responsible for developing the process to the condition, essentially as used today, as the most common of the nickel refining methods.[15] All of the available methods of refining nickel are from the International Nickel stable.

Although the salts business was continued for several years, in 1905 Wiggin decided to change direction and become a supplier of semi-finished alloy products. It had previously supplied ingot to alloy processors and, for those customers who required flat products, the rolling had been carried out by Messrs. Rollason. In 1905 Rollason's Bromford mills were acquired and the equipment removed to Wiggin Street. The initial equipment was for cold strip rolling only, but hot rolling and rod rolling for wire production soon followed.

The post 1914–1918 war period had the same effect on the nickel industry as other metal industries. There was a lot of scrap metal available from the recycling of military equipment, much of it containing significant levels of nickel. The economic weakness that followed the termination of hostilities, together with the availability of military scrap, resulted in a poor market for both Wiggin's and Mond Nickel's products. This resulted in them merging their activities in 1919; Mond acquiring the entire share capital of Wiggin. Mond-Nickel had established a rolling mill in Birmingham just before the merger, signalling an extension of the competition with the Wiggin business, but the semi-finished products business was developed subsequently on the Wiggin Street site.

International Nickel had been formed in 1902 by the merger of the Canadian Copper Company, a mine operator, with the Orford Copper and Nickel company and for a period of 24 years Mond Nickel and International Nickel were in competition for the market for the nickel they were both extracting from the Sudbury Basin. Both companies were dedicated to researching new ways of using nickel to increase the market for their products, and for both the urgency of finding new outlets increased with the decline in business that followed the 1914-18 war. Howard-White notes that the Mond activity was scientifically led, while that of International Nickel was more market orientated.

For many years International Nickel restricted its activity to mining and refining, to which it later returned. It developed only one alloy in its early years, as recounted in Chapter 10. It was of interest to a Scottish company, which needed strong corrosion resistant alloys, The Weir Group.

Lord Weir visited International Nickel in 1910 and persuaded the company to give him a licence to produce the alloy, Monel, for use in their business. Since International Nickel was only a miner with limited refining capability, a committed producer of one of its alloys gave it a tied market. Weir built a plant at Thornliebank near Glasgow, calling the company Monel-Weir.[16] There were no corrosion resistant steels available until some years later but by the 1930s stainless steels were widely produced and the Monel business was of less interest to Weir. The plant that had been established was sold to International Nickel in 1932 and placed under the management of Henry Wiggin & Co.[9]

(At the time that the Monel-Weir business was transferred to Wiggin, The Weir Group became International Nickel's agent in France, where Le Nickel is based. International Nickel had an agreement not to place its own agents there).

After Mond acquired Wiggin, the British research activity was located at Wiggin Street, where it remained until 1982. In the USA, research was at first located at the Orford works in Bayonne, New Jersey. International Nickel also established a rolling mill and entered the market for semi-finished products after the 1914-18 war. Development work was carried out there, in Huntington, West Virginia, and at the laboratories that the company established in New York state and in Canada.

Following the purchase by International Nickel of the Mond Nickel Company in 1926, the research activity in Birmingham was expanded.

In 1928 the company acquired the properties along the whole of Wiggin Street. The premises shown in Figure 2.4 included in 1887, a galvanising works, located in the premises shown in Figure 2.3, originally occupied by the British White Lead Company.[17] Having acquired the whole of the site between the canals, Wiggin Street and Icknield Port Road, the first development was the building of a smart

Figure 2.6 *The Development and Research Laboratory in 1936. (Courtesy Peter Crouch)*[18]

new research centre, shown in Figure 2.6. This was located towards the canal end of Wiggin Street.

The remainder of the site was developed with additional and replacement equipment for melting and casting, forging and rolling and wire drawing. The business of Henry Wiggin continued there until 1960. Thereafter all the Wiggin activities were transferred to a new factory in Hereford. Clydach continued to be a source of refined nickel.

The only activity left on the site after 1960 was research. Rather than move the research to Hereford it was decided to expand the activity at Wiggin Street and a major redevelopment deployed the mill buildings to supporting research and a new office block was added at the front. At this time, there were almost 300 people engaged in the research and development. Many are shown in a picture of the site from 1973 in Figure 2.7.

This was an ill-fated development. Research continued in Birmingham until 1982. Until almost this time, International Nickel had dominated the world market for nickel and had continued its search for new ways of using the metal that it could share with the world. It made sense for the company to offer the fruits of its research to the world, knowing that it would benefit from enhanced demand and higher prices. However, by the 1980s there were several competitors sharing the demand and INCO's research was effectively subsidising its competitors. Inevitably the company's commitment to research into the uses of the metal weakened and it chose to concentrate on process research, with the aim of being better at producing the metal. This it continues to do at a laboratory near Toronto, with good effect.

The business of making products from nickel and its alloys also became less valuable to the company as there were many more companies involved, all of whom were potential customers for INCO's metals. All the downstream businesses were sold in 1997.

Others may claim that the new metals, titanium and zirconium, are the most characteristic of Birmingham's metallurgical tradition, but the Government-sponsored development of these metals was in imitation of the US developments in Oregon. Within half a century it was absorbed into the development of the more successful American

Figure 2.7 *Research staff in front of the new offices in 1973.*[18]

industry. The nickel industry not only survived there for a century and a half, it truly started in Birmingham.

In the following chapters the many ways in which nickel, in combination with other materials, has changed the world will be described, linking each application to the reasons for its effect and to the evolving understanding of the forces that govern material behaviour. The treatment will follow a roughly chronological order. It is hoped that explanations given of how nickel acts to promote the many applications will be comprehensible to the non-expert but rarely offend the experts.

3
One of Many

There are fewer than a hundred naturally occurring elements. Nickel is just one of them, but it has found its way into many applications that we find necessary. In this chapter the objective is to illustrate and explain how it was formed and how it compares with the other elements. Only the general principles of the properties of materials will be described. The characteristics of nickel are more fully examined in later chapters.

Since Albert Einstein formulated the theory of relativity and developed the deceptively simple equation:

$$e = mc^2$$

we have known that everything in the Universe is a form of energy. Little packets of energy with various properties combine to form positively charged protons and negatively charged electrons and some of these combine and form electrically neutral neutrons. Others, together with additions of neutrons form atoms, which are thought of as microscopic solar systems, with the protons and neutrons at the core and the electrons orbiting around them. The simplest atom, hydrogen has one proton with one electron. The next one, helium, needs some neutrons at its core, because it has two protons which do not like to be too close together; like charges repel, opposite charges attract.

Atoms like hydrogen and helium can be forced together with high pressures and temperatures, producing further atomic species, lithium, beryllium, boron, carbon, nitrogen, oxygen and so on by nuclear fusion, which can continue at increasing pressures and temperatures, but still generating energy in the process of fusion until an isotope of nickel, Ni^{62} is reached. This is the most stable isotope; Fe^{56} is a little less stable. Higher weight isotopes are less stable, and many are not stable at all but decay, that is break into fragments with the emission of radiation.

The stability of the isotopes varies as shown in Figure 3.1.

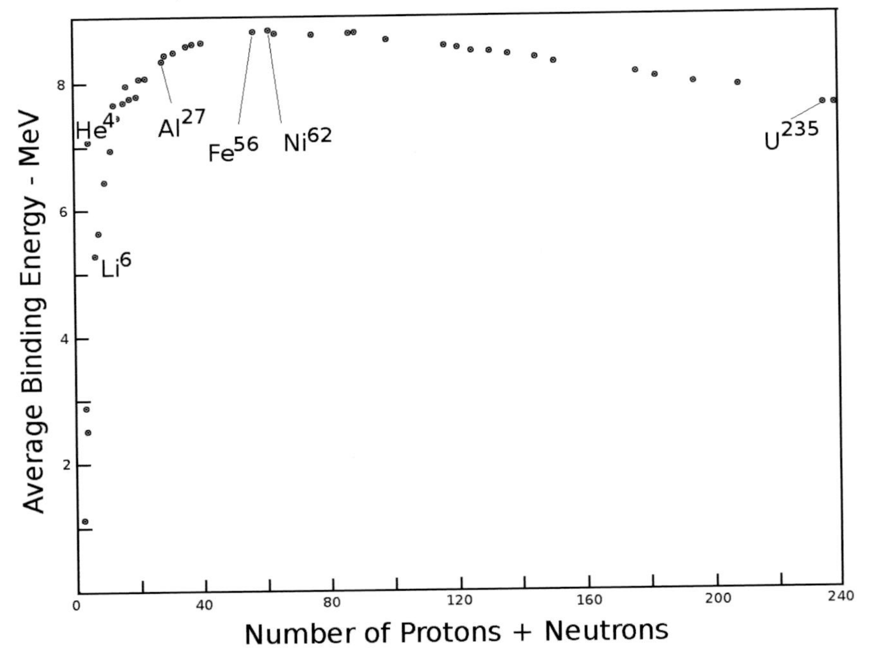

Figure 3.1 *The nuclear binding energy as a function of atomic weight.*

The greater the binding energy between protons and neutrons at the core of the atom, the more stable the isotope. It is thought that the reason why the core of the earth is mainly iron, and nickel is because of the stability of their isotopes.

Interatomic bonding typically increases with the atomic weight, as would be expected from the increasing numbers of electrons and protons. This can be inferred from the density, which is a result of the atomic size, how strongly the atoms pull together and how they pack together. This is shown for the solid elements in Figure 3.2. The rough line is the average atomic weight of the elements. The elements above

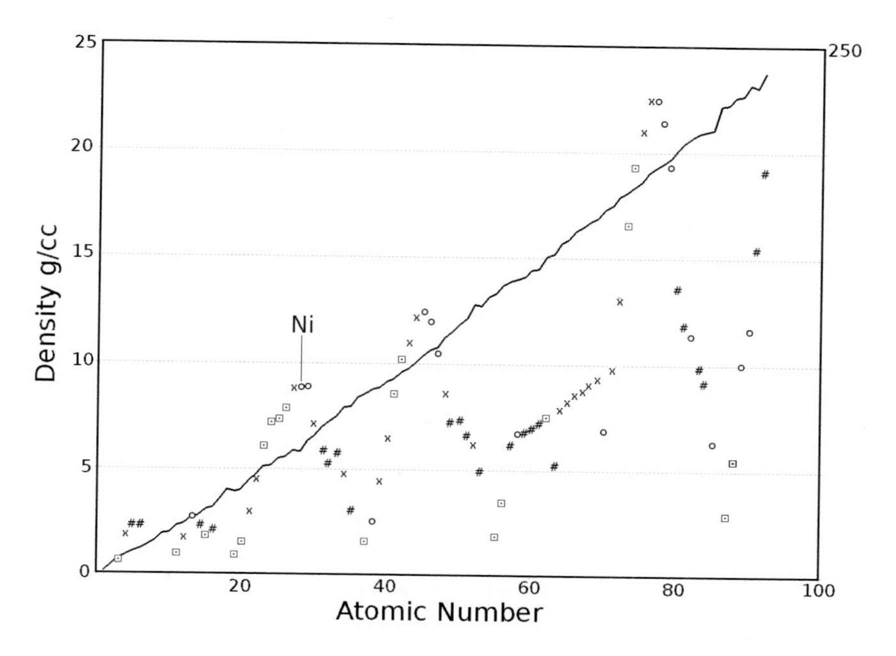

Figure 3.2 *Density of the elements at room temperature as a function of atomic number.*

the trend line are more strongly bonded, more closely packed, or have smaller atomic sizes for their atomic weight. These comprise mainly the transition metals, including nickel, which also bonds strongly to other transition metals.

An obvious feature contributing to the density of an element is the crystal structure that it adopts. There are three common ones and several less common types. The common ones are: face centred cubic, the elements having this structure are as circles; body centred cubic, shown as squares; and hexagonal shown by x. Others are shown as #. Assuming that the atoms act like hard spheres, in the fcc structure they would occupy 74% of the volume. An ideal hexagonal structure would have the same packing density, but the planar spacing is rarely close to the ideal. In the bcc structure 68% of the volume would be filled.

The crystal structures are illustrated in Figure 3.3.

There is an obvious effect of atomic weight on density, but the fact that the strong metals lie above the trend line indicates that they also have a stronger interatomic bond than the other elements. A further indication of this can be seen in the elastic moduli of the elements, represented in

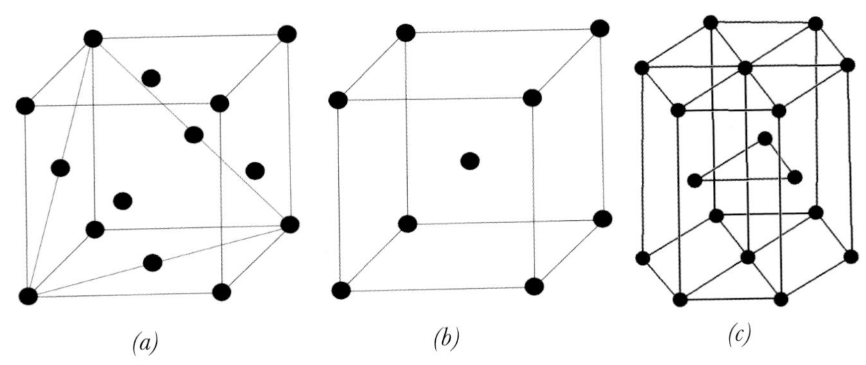

Figure 3.3 *a) the face centred cubic structure, b) body centred cubic and c) hexagonal.*

Figure 3.4 by Young's modulus. That is the ratio of the stress to the strain it produces when loaded in tension. It is a measure of the difficulty of pulling atoms apart. The very high modulus is for the diamond form of carbon. This remarkable element can adopt several structural forms. The very soft form, graphite, is very strong in one plane, but the planes are weakly connected, so that graphite has lubricant properties.

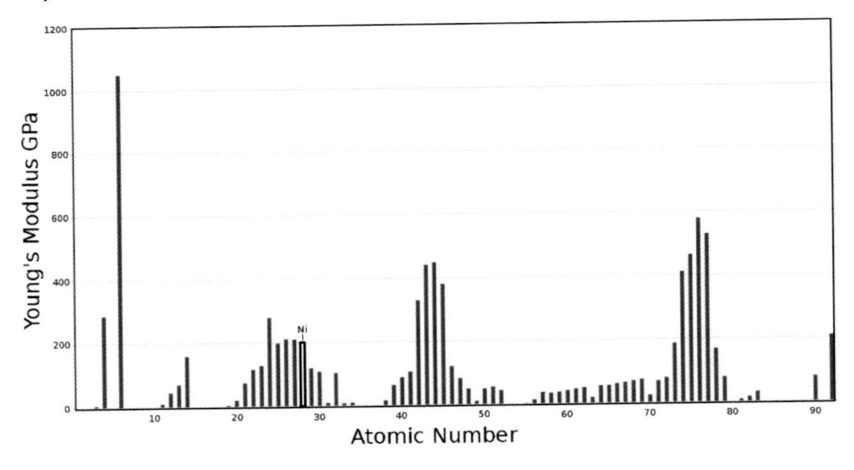

Figure 3.4 *Young's modulus of the elements.*

To get some idea of what is causing the different metals to have different bond strength it is necessary to appreciate more about the structure of the atoms.

The similarity of chemical behaviour of some of the elements alerted several workers to periodicity among them. The first to produce a periodic table was John Newlands (1837–1898). He arranged 57 elements into eight groups in 1865. Interestingly, Newlands did not recognise that cobalt and nickel were distinct elements and he grouped them together as element 22. Mendeleev (1834–1907) improved on the arrangement in 1869 and accurately predicted that elements as yet unknown would occupy gaps in the table he produced. His periodic table is reproduced in Figure 3.5.[1]

Like Newlands before him, Mendeleev could not separate nickel and cobalt. His ranking was based on the average atomic weight and, with cobalt weighing in at 58.93 and nickel at 58.69, on that basis they would not have been separable. This was despite an industrial process to separate the two having been in operation in Birmingham for some thirty years.

It was not until Henry Moseley (1887–1915) analysed their X-ray emission spectra that it was confirmed that the sequence of the elements is based on the number of positive charges each contains in their nucleus. It was then possible to complete the periodic table, subsequently adding only rhenium of the stable elements, discovered in 1925. Figure 3.6 is a subsequent format.

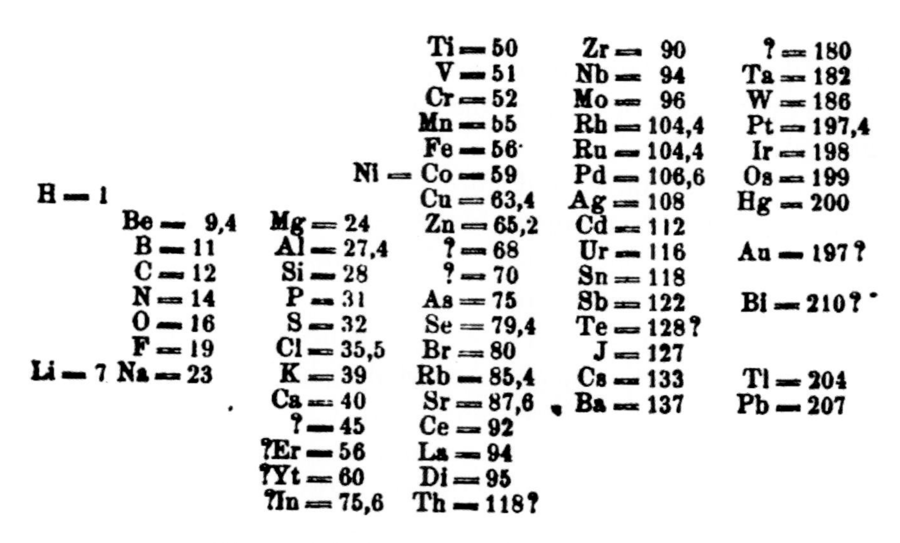

			Ti = 50	Zr = 90	? = 180
			V = 51	Nb = 94	Ta = 182
			Cr = 52	Mo = 96	W = 186
			Mn = 55	Rh = 104,4	Pt = 197,4
			Fe = 56	Ru = 104,4	Ir = 198
			Ni — Co = 59	Pd = 106,6	Os = 199
H = 1			Cu = 63,4	Ag = 108	Hg = 200
	Be = 9,4	Mg = 24	Zn = 65,2	Cd = 112	
	B = 11	Al = 27,4	? = 68	Ur = 116	Au = 197?
	C = 12	Si = 28	? = 70	Sn = 118	
	N = 14	P = 31	As = 75	Sb = 122	Bi = 210?
	O = 16	S = 32	Se = 79,4	Te = 128?	
	F = 19	Cl = 35,5	Br = 80	J = 127	
Li = 7	Na = 23	K = 39	Rb = 85,4	Cs = 133	Tl = 204
		Ca = 40	Sr = 87,6	Ba = 137	Pb = 207
		? = 45	Ce = 92		
		?Er = 56	La = 94		
		?Yt = 60	Di = 95		
		?In = 75,6	Th = 118?		

Figure 3.5 *Mendeleev's original periodic table.*

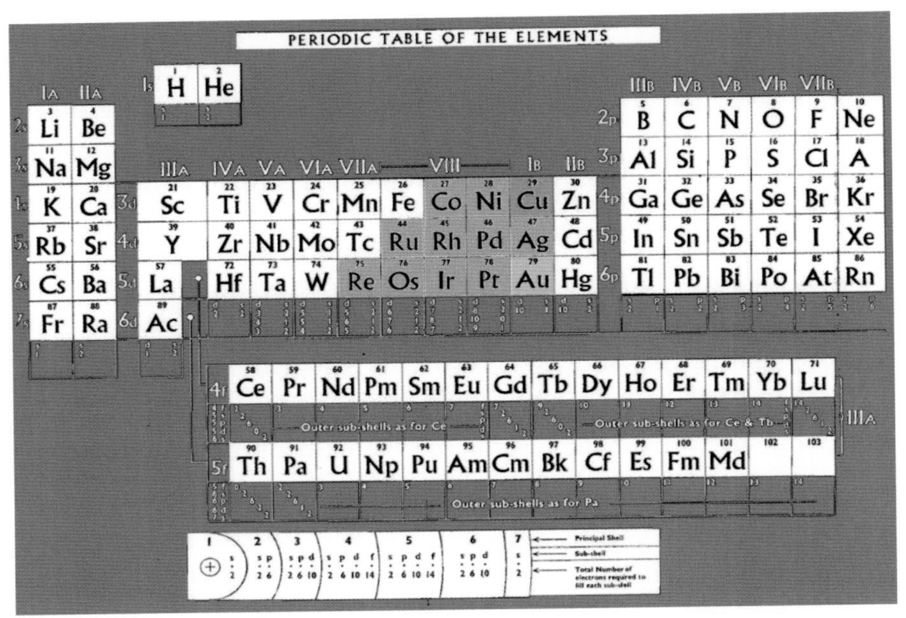

Figure 3.6 *Periodic table derived from the version supplied by International Nickel to schools and Universities for many years.*

Having established the order in the table, and with the important discovery of the inert gases, helium, neon, argon, krypton and xenon in 1894 by William Ramsay (1852–1916) and Lord Rayleigh (1842–1919), the interpretation of atomic structure became clearer.

An explanation of the periodic table is given in the first section of the Appendix.

The periodic table in Figure 3.6 shows the various electron configurations of the elements. In the first three rows, passing from left to right, the s electrons first advance to the maximum, 2; and then the p electrons are added in sequence in rows two and three. In row 4 and again in row 5 there are now d electrons, which are added before the p electrons, but now the energy levels of the s and d electrons are close together and several of these transition metals can assume more than one electron configuration. Most importantly the d electrons can participate in the formation of bonds between atoms. The higher density and relatively higher Young's moduli of the transition metals are the result, explaining at a simple level why most of the engineering materials are among the transition metals.

The generally rising density and moduli from the transition metals in row 4 to row 5 to row 6 can be viewed as a simple matter of electrostatics. The interaction between two like atoms approaching each other from a distance is illustrated schematically in Figure 3.7.

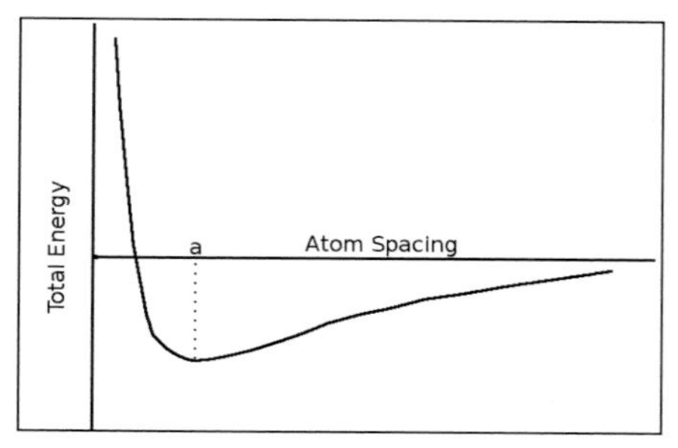

Figure 3.7 *The interaction energy between two atoms as a function of their separation.*

The concentrated positive charge of the nucleus is screened at long range by the distributed charge of the orbiting electrons, so that the positive charge of each atom attracts the negative charge of the other. When the two sets of orbiting electrons start to interact, they repel each other, and the nuclei also react against getting closer. The moderate long-range positive interaction is countered by a strong and steeply rising short range repulsion. The forces balance out at the spacing **a**, which is the position of minimum energy.

The density and moduli increase with increasing atomic number for the transition elements because the electrostatic forces are proportional to the charges that each has.

Figure 3.6 has been shaded to show that nickel and cobalt are members of a group in the periodic table that include the precious metals. These are all metals that are relatively unreactive, especially with oxygen, but also with carbon and nitrogen. The s electrons are the main agents of bonding and to be reactive a material must have free bonds at its surface. The ability to adopt different electronic arrays provides a simple explanation for why these metals are noble. Having a different surface configuration with fewer bonds than at the interior of a crystal deprives corrosive elements of any purchase.

The corrosion resistance of nickel, especially its resistance to oxidation, underlies many of its applications and will be met with in many of those that are to be discussed. However, like several of the noble metals it does have interesting reactions with hydrogen, as will be discussed later.

A further insight into the nature of the metallic bond is given by the fact that the one characteristic that truly defines a metal is its conductivity. Thermal and electrical conductivity have a lot in common and it is the thermal conductivity that is shown in Figure 3.8.[2]

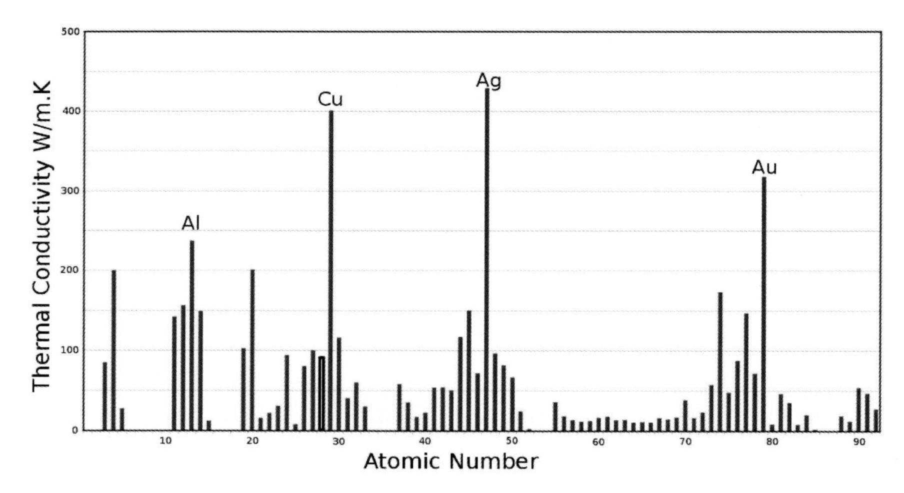

Figure 3.8 *Thermal conductivity of the elements.*

The main difference between electrical and thermal conductivity is that in solids at low temperatures electrical conductivity is solely due to electron migration, whereas thermal conductivity also has a contribution from general lattice vibrations, thus all solids will show some thermal conductivity. Metals are best, except for the diamond form of carbon (gases and liquids transfer heat by convection and conduction through the movement of their molecules). Gold, silver and copper are the most conductive metals. Aluminium is the next best, but nickel has quite poor conductivity.

There are several models for the structure of a metal. The simplest is that a metal crystal is an array of positive ions which shares the bonding electrons between them, so that no valence electron is attached to one metal ion. This is the band theory of metals, which has the electrons occupying a range of energies. This is a useful theory for understanding behaviour when elements are mixed to make alloys. Another theory, due to Linus Pauling (1901–1994), postulates that the bond between atoms is made by the local sharing of electrons, which resonate between positions on adjacent atoms.[3] Such electrons would be more restricted in their movement.

The evidence of the conductivity suggest that the truth may be that these two models represent a spectrum rather than being competitors for the truth. The three most highly conductive elements, silver, copper and gold, all have one s electron. If this was shared with all other atoms, the electrons would have no difficulty in moving to contribute to the electrical current or heat transfer. Aluminium, to a lesser extent, may tend towards the free electron theory. But most of the transition metals are much poorer conductors, suggesting that the electrons leave their location less readily.

Although not a good conductor, the properties of nickel and its alloys are important in electrical applications as will be seen later. Some of these depend on another property it shares with very few other elements, its ferromagnetism. All elements display a magnetic response, but only iron, cobalt, nickel and some rare-earth metals at very low temperatures, react strongly to the presence of a magnetic field. These are the ferromagnetic materials.

Magnetism is discussed further in Chapter 12 and Appendix 1.2. It is a property that can only be understood through a familiarity with quantum mechanics. The simple explanation is that it derives from the spin of electrons. The notion is that each electron attached to an atom occupies a unique energy level defined by four parameters. The first three parameters derive from an explanation of the layout of the periodic table; the pattern can be deduced from the structure of the inert gases. Helium, having two electrons, needs two energy levels. Although the absolute energy is the same for both, the two electrons have their fourth parameter, called their spin, in opposite directions. Neon has 8 electrons, consisting of four pairs with opposite spin, and so on.

To understand magnetism, it is postulated that the spin on each electron makes it into a tiny magnet. If the electron states were filled so that the electrons are paired, with both senses of spin being represented at each energy level, or, for elements with an odd atomic number, the atoms correlate to give no overall tendency to one direction of spin, the material will not be magnetic. It is therefore implicit that the atoms of ferro-magnetic materials prefer one direction of spin. This is the case for iron, cobalt and nickel.

In the demagnetised state, these metals are composed of little magnets at a microstructural level, but they are randomly oriented so that the magnetism cancels out overall. When magnetised to saturation all the atoms have aligned their unpaired spins. Many compounds of these metals are also ferromagnetic as also are compounds of manganese. It is, in the first long period, the spin on the d electrons that prefer to choose one direction rather than the other, so the magnetic effect varies with the average number of d electrons an alloy possesses. This is illustrated in Figure 3.9.

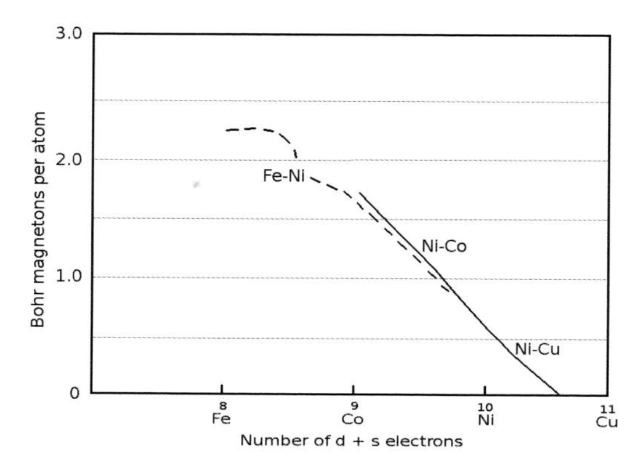

Figure 3.9 *Maximum magnetic moment as a function of the number of d + s electrons. (After Bozorth).*[4]

In Figure 3.9. the Bohr magneton is a measure of the maximum magnetism that the metal can have. Nickel and cobalt behave according to the rule of mixing. The metals mix freely when they have the same crystal structure. Iron and nickel show an anomaly, which is because they do not have the same crystal structure at room temperature. This will be met again in Chapter 7.

It has already been noted that nickel is resistant to oxidation, a property that underlies many of its applications. It has a greater affinity with sulphur. One of its main sources is in sulphide ores, although the greater quantity are weathered silicate minerals. The abundance of these in the earth's crust is one determinant of its price and availability. It is far from the most abundant of elements, coming in at 23rd in the list.[2] Almost half of the crust consists of oxygen, found in oxide, hydroxide, carbonate, silicate and other minerals.

Several more abundant elements find few applications. For instance, barium, the 14th most abundant at 500 parts per million and strontium, 16th at 370ppm, may only be familiar to people who have experienced specific medical diagnostic techniques. Most of the more abundant elements are highly reactive ones, therefore much more difficult to separate from their ores than nickel and cobalt.

Some of these elements will become more important as economic methods of production become available and their properties known better. Titanium and zirconium are perhaps the two most likely to find wider uses, titanium is ninth most abundant and zirconium the 18th. Both have excellent low temperature properties. They would be more widely used if better means of separating the metals from their compounds were available.

Nickel is so important, as will be shown in the following pages, because it was made available early in the process of industrialisation as a pure element; that it has a range of properties that is only equalled by its neighbour, cobalt and it is adequately available, more so than cobalt, in the earth's crust.

4

Better than Silver

The initial use of nickel in artefacts was a consequence of its appearance and the fact that the natural alloys did not tarnish or corrode. They would initially have seemed superior to the only other white metal that was widely known, silver. When made available in Europe the white cupro-nickel alloys soon displaced silver from the cutlery trade, a trend that was strengthened when it was found that the alloy was the best substrate on which to plate silver to create a cheaper product that was superficially like the silver ware of the gentry. Electroplated nickel-silver was the dominant material in this business for all but knife blades until it was displaced by stainless steel in the second half of the twentieth century.

As noted in the historical review, German-silver was first used for table ware. Cutlery, tea and coffee pots, table illumination using candelabra had all been made from solid silver for the rich for centuries. The development of Sheffield plate by Thomas Boulsover (1705–1788), produced by laminating thin layers of silver onto a copper base, had extended the market for goods that mimicked solid silver to the next level of society from the second half of the eighteenth century. After German-silver came available it was found to bond well with silver and it mattered less if the silver wore off, so thinner layers of silver could be used.

Leaving the silver off altogether was a further way of reducing the cost, extending the market for silver-coloured metal utensils further down the social scale. Since the un-silvered material is functionally superior, below stairs, even the nobility had tools and utensils that were made of nickel silver, as exemplified by the serving spoons and fork shown in Figure 4.1.

Figure 4.1 *Two nickel-silver table spoons and a fork, with their maker's marks.*

Of the three items in Figure 4.1, the two spoons were produced in Birmingham, the fork in Sheffield. All are a hundred or more years old, having already served five generations of a family. The upper spoon is from a company founded late in the nineteenth century, Daniel and Arter, who used a variety of names, including the Bengal Silver seen here. The fancy name was probably given because it is a poor grade. Bengal silver was one of several names used by the company.[1] The metal has a yellow tinge, indicating that it is low in nickel. The other spoon was made by the firm of G.R. Collis, which had been started by Edward Thomason. He learned his trade at the Soho Foundry, where he was articled in 1785. The fork is from Viners, having their brand name Resilco stamped on the handle.

The usual composition of nickel-silver is approximately 4 parts by weight of 70/30 brass to one part of nickel. This is enough nickel to change the metal colour from yellow to silver. Brass founders commonly added German-silver to their range of products, its name surviving as the common appellation until the early part of the twentieth century, when anything German was unpopular.

The predominance of the brass constituents, copper and zinc, perhaps justifies most bands that use only metal instruments in being called brass bands. Some are indeed referred to as silver bands, but the big players are the ones that participate every year in the National Brass Band Competition.

One of the very oldest, originally called the Black Dyke Mills Brass Band, features in Figure 4.2. It is still winning championships, having been formed in 1855. As can be seen the instrumentalists play silver coloured instruments. Today they are nickel silver, but the record shows that the band was formed using instruments supplied by Joseph Higham of Manchester.[2] There is a detailed record of the manufacture of these instruments, where it is clearly explained that the first set was made of a special, but undisclosed, brass and were silver plated.[3] That was only a few years after Elkington patented the plating process.

Figure 4.2 *The Black Dyke Brass Band. (Courtesy of the Band)*

Concert orchestras also feature significant quantities of nickel-silver. It is used for the metal flutes, and the keys of clarinets, oboes and bassoons, while jazz and dance bands feature saxophones of various shapes and sizes made in the metal.

Musical performers did not wait for nickel-silver to be available before making music, but a significant development in the last quarter of the nineteenth century did benefit significantly from its availability and an unusual property of the alloy.

Like most metals, nickel-silver develops a thin oxide coating. For most metals and alloys, the oxide layer is an insulator, so that if a metal component is to be part of an electrical circuit it is often gold plated. That works well in a situation where once the connection is made it is rarely broken. Where a gold-plated device, such as a micro-SD card, is repeatedly inserted and removed, it doesn't last long. The soft gold wears off. For the weak currents in computer applications that must be tolerated. For more robust applications nickel-silver can be used, because its oxide is a conductor.

The most important application that came to depend on the alloy was in the early days of telephony. The first telephone exchange was developed by Tividar Puskás (1844–1893), who was at the time working for Thomas Edison (1847–1931). Edison's graphite microphone and Bell's telephone patent had made point-to-point voice transmission possible. What was now needed was a means of linking any two points. The first manual exchange was opened in 1878 in New Haven with 21 subscribers. The idea spread quickly, as illustrated in Figure 4.3.

The principle used is obvious. It involved audio jacks connected to the caller, which were pushed into sockets connecting to the called. Both jack and socket were made of nickel-silver.

Figure 4.3 *A telephone exchange from 1892. (Wikimedia Commons)*

Workers losing their jobs to automation is not a new phenomenon; the example of telephony shows that it goes back a long way, in this case to the first half of the twentieth century. None of the jobs of the women making the connections in Figure 4.3 survived the introduction of automatic telephone exchanges.

The earliest automatic exchanges were electromechanical. One example from an AT&T video of the mechanism in use by Bell Telephone in 1951 can be seen on YouTube.[4] These continued to use nickel-silver for the contacting parts.

Each of the connections required a series of relays to direct the electrical pulses along the specific route that connected the caller and the called. Each relay had to make the connection cleanly; the mating surfaces had to be robust and durable and the material used was still nickel-silver.

Later exchanges were only electrical, using the semi-conduction capability provided by diodes, which will be encountered later; while modern systems are built on the materials we call semi-conductors and are described as digital systems.

The market for nickel-silver has been substantially reduced by newer materials, but the combination of conductivity and endurance still finds application in one very familiar hobby, as shown in Figure 4.4. Here the electrically conducting components are best made of nickel-silver.

Figure 4.4 *A model railway layout. (Courtesy Michael Dillamore)*

5

Making Money

There has been nickel in coins since long before the element was isolated but designing coinage around the properties of nickel and its alloys clearly was not possible before it was widely available in elemental form, so its usage became widespread in the second half of the nineteenth century.

Nickel has been used in coinage for two essential reasons. It has the property of being miscible with the red and yellow metals, enabling the alloy colour to be tailored to signify value and it confers resistance to tarnishing that betokens the excellent corrosion resistance that it confers in all of its applications. As this was the first application where corrosion resistance was a required attribute, the background reasons for this property will be introduced here, although its importance arises frequently in much more demanding areas that will be discussed later.

A company brochure issued by the Mond Nickel Company records that, by 1918, 74 countries and overseas territories of the colonial powers were using unalloyed nickel in coinage. The brochure has a map of the world showing how widespread it was at that time. It is reproduced in Figure 5.1.

Cupro-nickel coins were also in use, although not in the UK until 1947. In America the five-cent nickel, which is actually cupro-nickel, has been in use since 1866.

Colour and durability are the key considerations for metallic forms of money. Copper when mixed with various other elements can have a

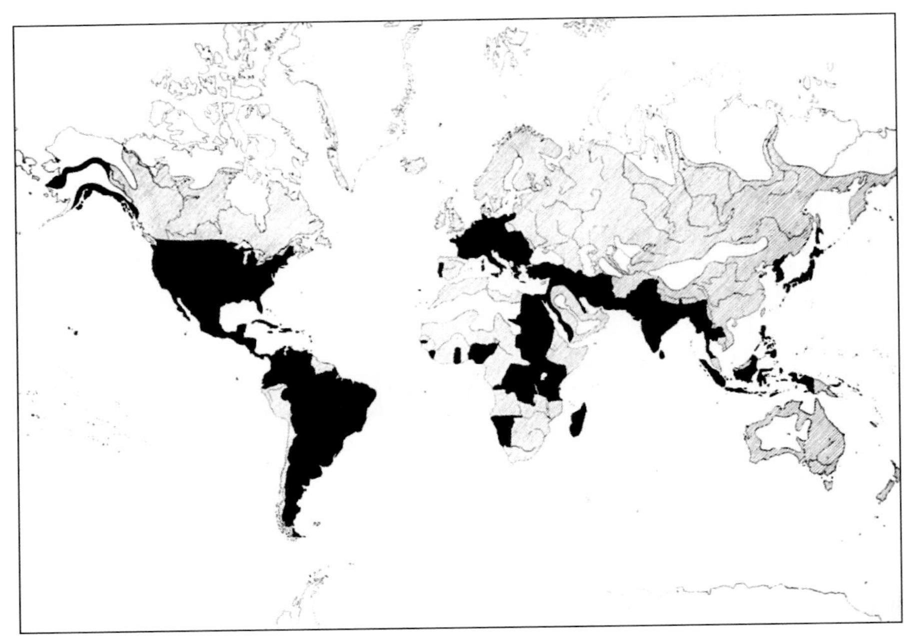

Figure 5.1 *World map showing, in black, the areas where unalloyed nickel coins were in use by 1918.*[1]

wide range of different colours in a spectrum from white to red. Three variants are used in British coins, as shown in Figure 5.2.

The two pence piece is shown as being mostly copper with just 2.5% of an alloying element, in fact, tin, not zinc. The tin addition is added to make the metal a little harder – pure copper is too soft to be used in this application – but it has little effect on the reddish colour of copper at this low concentration. Additions of both zinc and nickel to the copper progressively dilute the copper colour. Only about 15% of nickel is enough to produce a white alloy; for zinc, almost 50% is needed, but at this level the alloy is brittle and is of no interest. Only five compositions, as shown, are used for the various denominations of British coins. Until 1992 both the 1p and 2p coins were made of the dilute bronze alloy; new 1p and 2p pieces since

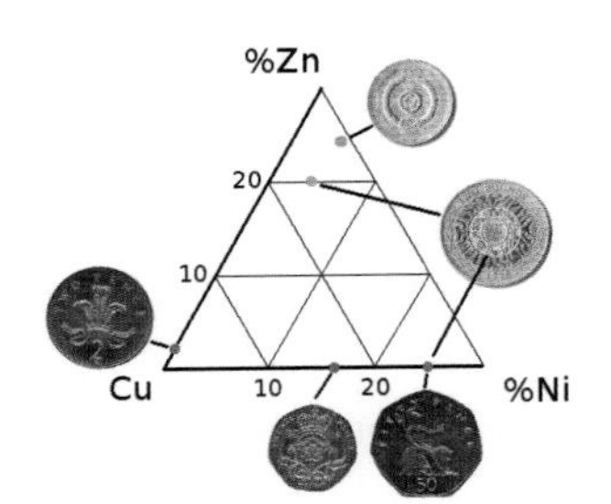

Figure 5.2
The compositions of Cu-Zn-Ni alloys used in UK coinage.

1992 have been made of steel plated with copper. Old 5p, 10p, 50p and the inner part of the £2 coin are 75% Cu and 25% Ni. The 20p coin was also a simple cupro-nickel but containing only 16% of nickel.

The Royal Mint has now introduced steel coins plated with nickel for the 5, 10, 20 and 50p denominations. Whether this continues to be promoted will depend on how many people claim to be sensitised to the well known nickel-contact allergy.

With the banks hoping to eliminate coinage and paper money, the demand for metal in coinage is now diminishing.

The alloy used for the £1 coin has slightly more nickel and zinc than the outer segment of the £2 coin, producing very similar shades of yellow. These alloys all have the same face centred cubic crystal structure as pure copper.

It is of interest that the colours chosen for modern coinage, which, in fact, consists of tokens that have a face value unrelated to their intrinsic worth, are chosen to mimic the colours of the earlier currencies. Intrinsic value was then related to face value. The commoner copper was used for small coins, silver for those of greater value and gold for the most valuable.

Apart from slight differences in the appearance in the range grey to silver, most metals have similar colours. This is because light in the optical spectrum does not interact with the electrons in the surface of the metal. The differences in the grey to silver range are mainly due to the thin oxide coating that is usually present, although there may be some interaction at the blue extreme of the spectrum and both infra-red and ultra violet light do interact with the metal. Infra-red causes warming, which simply means that the energy in the incident wavelength causes the atoms to vibrate more. Ultra-violet light can stimulate conductivity by promoting electrons into the conduction band. However, for the grey metals the main interaction is to reflect the light. If the metal is highly polished it acts as a mirror, reflecting incident light normally. For common artefacts, the surface is not smooth, and the light is scattered so that the object is less bright, i.e. grey.

The peculiarity of copper and gold is that they do interact with light in the visible spectrum, so that light of wavelength of about 600 nanometres and less is absorbed by copper and light of 570 nanometres and less is absorbed by gold. The photons of these wavelengths have energies greater than 1.9eV for copper and 2.1eV for gold approximately. The key to the interaction is found in the electronic structure of the metals. The periodic table of the elements shown in Figure 3.6 has the relevant information. It has been shaded to indicate the metals that are most resistant to oxidation.

The group Ib elements copper, silver and gold, are all shown with the d-band having the full complement of 10 electrons, with just one s electron. Earlier in the long periods preceding these metals it is most common for there to be fewer than 10 d electrons, but the maximum number, 2, of s electrons. Each electron attached to an atom has a unique energy state, and an atom can only absorb incident energy if it has a vacant energy state into which an electron can be promoted. Since the group Ib elements do not have 2 s electrons they fulfil this requirement. However, it is also necessary that the incident photon energy is great enough to raise the energy of a d electron into the s-state. This condition can evidently be met for copper with wavelengths shorter than orange, for gold by a slightly higher energy, but for silver the requisite energy is at least high in the blue range.

A simple band theory can reasonably explain what happens when the pure metals are mixed with metals that contribute additional valence electrons. This envisages that the additional valence electrons occupy increasingly high energy states, so that the energy gap between the available d electrons and the top of the energy band increases, requiring higher energy photons to effect the transfer. The spectral range that is absorbed decreases as the alloy composition deviates further from the pure metal and the colour of the alloys graduates from the yellow and orange towards silver.

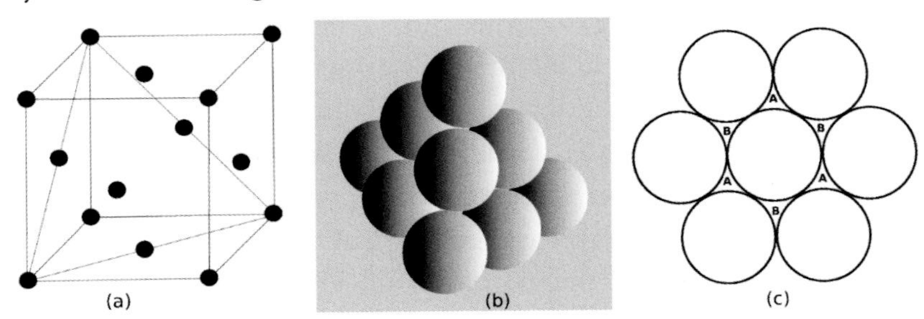

Figure 5.3 *The atomic arrangement in the face centred cubic crystal structure.*

All combinations of gold and copper have colours in the orange to yellow range. For coinage and the other early uses of nickel alloys, the main attraction was that they resist corrosion.

All noble metals are also face centred cubic (fcc), but some very reactive metals, such as the alkaline earths, some of the rare earths and aluminium, lead and thorium also have this structure. For the noble metals the ability to vary their valency is important.

Within a crystal each atom in the fcc structure is surrounded by 12 nearest neighbours, as may be inferred from Figure 5.3.

In Figure 5.3c it is evident that the central atom has six near neighbours in the hexagonal array plus the three at A on the plane above, and the three at B on the plane below. How an atom with two valence electrons forms bonds with its 12 near neighbours is often accounted for on the basis that each atom gives up all its valence electrons to a cloud surrounding the positive ions. An alternative description, due to Pauling[2], has the bonding described as of a resonating covalent nature; that is each atom moves its valence electrons around from neighbour to neighbour.

In the diagram, (a) shows the formal cubic arrangement with atoms at each corner and in the centre of each face of the cube. The triangular trace is one of four equivalent close-packed planes, which can be recognised in (b). In (c) the stacking of each of the close-packed planes is in a sequence of three positions. The plane above the hexagonal array has the atoms centred on the positions marked A; those on the plane below are aligned with the Bs.

It is the corrosion resistance of the nickel alloys that has established their place around the world as the preferred material for coinage, even giving the name 'nickel' to a unit of the most common of currencies.

A surface atom has fewer neighbours than one in the interior and the result is often described as leaving dangling bonds that are unsatisfied by near neighbour relationships. These unfulfilled bonds are thought of as enabling the metal to react with external, gaseous or aqueous chemicals to cause corrosion. The noble metals are distinguished, obviously, by being less corrodible than other metals and one possible contributor to this is evident in Figure 3.6, in that the noble metals can adopt different electron arrangements, reducing the number of valence electrons. Whatever the reason, the inclination of each element to form an oxide – the commonest corrosion product – is evidenced by the energy that is liberated when the oxide forms. This is illustrated in Figure 5.4 for several of the transition metals.

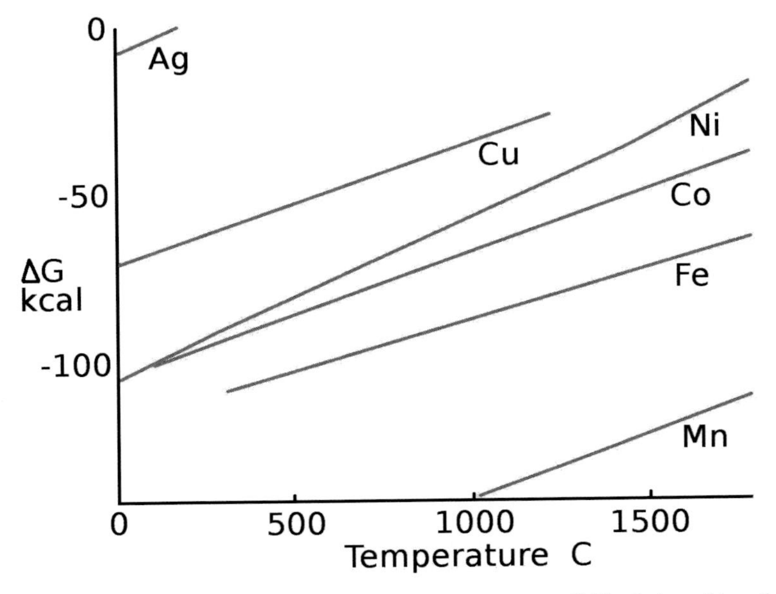

Figure 5.4 *The variation of the energy of formation (ΔG) of the oxides of some transition metals with temperature.*

The energy of formation is negative when the reaction is spontaneous, that is if it lowers the total energy of the system. The more negative the energy, the greater the affinity for oxygen, hence the higher in the graph the more noble the metal. Both copper and nickel can form oxides, which coat the surface of the metal, copper having a lower affinity for oxygen than nickel. However, nickel is more resistant to corrosion than copper because the oxygen that adheres to its surface forms a barrier to further access. From the diagram, there is a positive bond between a nickel and an oxygen atom, but the nickel is evidently not able, at ambient temperature, to break the oxygen bond of its molecular form. This is true also for copper.

At elevated temperatures both metals can be oxidised, but the oxide of nickel adheres to and covers the surface, while on copper the oxide layer thickens and spalls off.

A coating that forms to prevent further corrosion is said to passivate the metal. (This will be met with again in later chapters). The coating that does this needs several properties. The first was suggested in 1922 by Pilling and Bedworth.[3] Their idea was that the ratio of the volume occupied by a metal atom in the corrosion product to that occupied in the metal lattice, the oft-quoted Pilling-Bedworth ratio, has to be slightly greater than one. A lower volume corrosion product would not adequately cover the surface, a larger volume product would buckle and crack. Nickel, if forming NiO would have that ratio as 1.65, copper's is 1.78 for CuO as the corrosion product. The difference is not very significant, but the properties of the oxide, how resistant it is to cracking, will also be important.

The first corrosion product to coat the surface always creates at least a partial barrier to further attack; building to some thickness therefore requires diffusion of either the corroding element in, or the metal out, through the corroded layer. Both cases occur. How easily this diffusion occurs is evidently a further consideration.

For nickel, as for iron and cobalt, the presence of water increases the corrosion rate. This is evidenced by the green nickel hydroxide staining at the base of many electric kettles, which are nickel plated, (Figure 5.5).

Figure 5.5 *Nickel hydroxide staining in a kettle.*

As will be discussed later, iron, cobalt and nickel, and some of the noble metals, are able to break the hydrogen-hydrogen bond, taking hydrogen atoms into solution in the metal. The form of the water molecule helps in this.

The molecular weight of water is only 18, compared with 44 for carbon dioxide, yet at room temperature water is a liquid while CO^2 is a gas. The explanation for this lies in the shape of the water molecule, which is shown in Figure 5.6.

There is a residual electrical charge across the molecule, the hydrogen side being more electronegative than the oxygen, so that water molecules attract each. This is sometimes referred to as the hydrogen bridge. In contact with a clean metal surface, the negative, hydrogen side is attracted to the positive ions at the metal surface, giving the possibility of a hydrogen atom being absorbed into the metal.

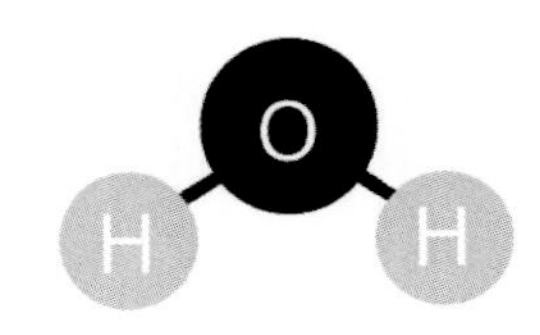

Figure 5.6 *Diagram of a water molecule.*

The hydroxide is readily converted to oxide by heating it to drive off a water molecule.

$$Ni(OH)_2 > NiO + H_2O$$

It is not only oxygen that nickel resists. Other elements that affect metals, importantly nitrogen and carbon, do not react with nickel. It is possible to form nickel nitride and nickel carbide, but only in very specific reactions that do not produce stable compounds. Iron, to the contrary, has strong reactions with both carbon and nitrogen. These are used in methods of hardening steel.

Its resistance to corrosion is one of the most important attributes of nickel; it is an ability that it takes with it into combinations with other elements. This is what makes nickel essential in most circumstances where a structural component carries a significant load in a corrosive environment. The materials developed, and their applications will form the major part of what follows. But the first major uses of nickel in nickel-silver and, as a result in coinage, were where this important attribute was recognised.

6

Electrochemistry

Electrochemistry is an important part of the nickel story in several respects. More nickel is now refined by electrolysis than by any other method: about 14% of all nickel is used to coat other materials to improve the corrosion resistance of artefacts, or simply to make them look more attractive: and nickel artefacts can be made directly by electrolysis.[1]

Electrochemistry was the start of the use of electricity. Static electricity had been understood by the end of the eighteenth century due to the work of Benjamin Franklin (1706–1790) and Joseph Priestley (1733–1804), but it was not until after Alessandro Volta (1745–1827) created the first battery that electricity could be supplied on demand, in a useful form.

Volta, an Italian, presented his discovery to the world in a communication, written in French, to the President of the Royal Society, Sir Joseph Banks.[2] The diagram from his paper, printed in Philosophical Transactions, is reproduced in Figure 6.1. The presence of electricity was recognised by putting each hand into one of the bowls at each end of the chain and experiencing a bigger shock as the number of cells increased. The shock was familiar from prior experience with static electricity.

The figure shows layers of the two metals separated by a wetted card. The aim of the diagram is to show that the effect of the metal pairs increased with increasing number.

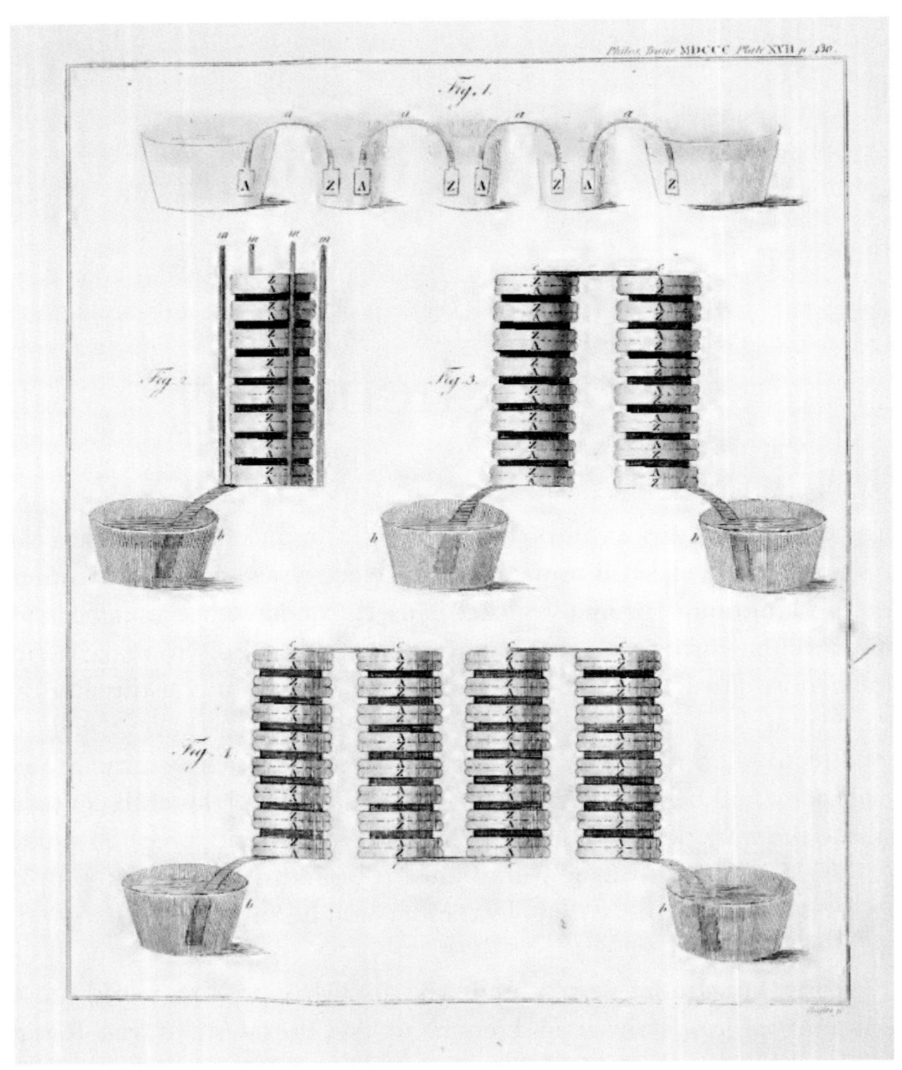

Figure 6.1 *Volta's piles. A, for Ag, is silver; Z for Zn, is zinc.*

Volta's experiments had been devised to refute the suggestion of Galvani (1737–1798) from his observation that a frog's leg twitched when touched with two wires, which he claimed was evidence that animals were a source of electricity. This had provoked much interest. One English researcher who had investigated Galvani's suggestion was William Nicholson (1753–1815) and Volta discusses his investigation in the Philosophical Transactions paper. Evidently Nicholson was converted to Volta's interpretation, because, in the same year (1800) he and Anthony Carlisle (1768–1840) recognised that the action of the current passing through the electrolyte had separated hydrogen and oxygen.[3] They thus discovered electrolysis.

Volta recognised that different combinations of metals produced different voltages and was responsible for the first electrochemical series. A partial representation of the modern data is shown in Figure 6.2.

The electrolyte used by Volta was a dilute brine. The less noble of the two electrodes, zinc, has a less negative electrode potential than sodium and could not therefore displace it from solution, so the positive ion displaced because of zinc supplying electrons is hydrogen. Had the electrolyte been a zinc salt, zinc ions would have been displaced from solution by the dissolving zinc and would have plated out on the silver electrode, the cathode. It is thus quite easy to coat a more noble metal with one that is less noble. This works in reverse if a less noble metal is immersed in a salt of a more noble one, for instance dipping iron into copper sulphate results in the iron being coated with copper and some iron ions going into solution. Henry Bessemer (1813–1898) was among the first to use this effect, when he coated some type-metal plaques he had cast, using a copper-containing solution.[4] The same principle was used for a time to coat artefacts with silver or gold, but the coatings are not very durable.

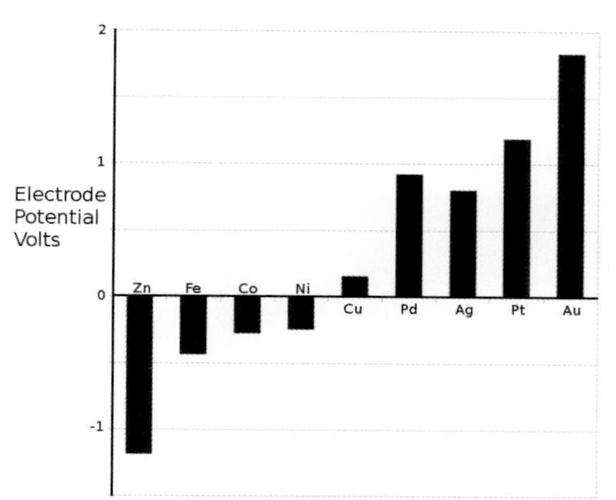

Figure 6.2 *Electrode potential of some common and some noble metals.*

To produce a better coating of the more noble metal it is only necessary to impose a voltage on the circuit, connecting the source and intended target, that exceeds the natural voltage produced by the couple. This is the basis of the, now widespread, business of electroplating.

However, electroplating did not become a serious business until after 1833, when Michael Faraday (1791–1867) showed that the amount of material transferred in electrolysis is proportional to the quantity of current passed. He proved that for a monovalent element, each electron that passes round the circuit releases one ion of the element into solution. For bivalent elements, two electrons must be released. However, this rationalisation had to wait until the nature of atoms and ions was better understood.

What Faraday's law did, together with Volta's rank order of electrochemical potential, was enable plating processes to be designed. This led very soon to the development of industrial electro-plating. After Volta and Nicholson and Carlisle, several researchers showed that one metal could be plated onto another.

In 1837 Bird showed that a variety of metals, including nickel, could be plated onto the cathode – in his case made of platinum – and this appears to be the first record of electroplating, although earlier workers had demonstrated that the action of electricity could decompose the oxides of metals.[5] It did not take long for the potential use of this to be recognised; the first British patent for nickel plating was granted to Joseph Shore[6] in 1840. He advocated using nickel nitrate as the electrolyte. Others followed using various solutions. One of the first in England to set up as nickel platers was the business of Bouse and

Figure 6.3 *Canning's advertisement in Peck's directory of 1896.*

Muncher. This was probably in 1870, when they started in business using a nickel ammonium sulphate solution that had been developed in America by Isaac Adams.

Continuing Birmingham's position as the centre of the nickel business, a few years later Cannings became established as the principal UK supplier of the chemicals and equipment required by platers.[6/7] The scope of the Cannings operation was illustrated in Peck's trade directory of 1896,[8/9] as shown in Figure 6.3.

Canning's essential ingredient, nickel, would have been mainly supplied by Evans and Askin, who also provided a limited range of nickel and cobalt salts.

Electrolysis is now the major method of refining nickel. The material produced by mineral processing is reduced using carbon, producing an impure form of nickel. This is then refined by electrolysis. There were several attempts to develop the process; success was eventually achieved by Hybinette in about 1900.[10] The impure nickel is made the anode in a cell in which the cathode is a material, to which nickel does not readily adhere, to make starter sheets. (Titanium is the current preference.) The impure nickel and the other impurities are all taken into solution and the purer nickel (sometimes with cobalt) plates out on the nickel starter cathode and the other metals are recovered in a sludge. Depending on the source of the original ore, the sludge may contain some precious metals, which are sometimes present in small quantities.

When the demand for nickel was at its height before and during the 1914-18 war, there were two refineries in the UK that produced nickel by electrolysis. The works of H.H. Vivian in South Wales was the bigger of the two, producing over 7 million pounds of nickel. The other one was at Erdington in Birmingham. Both used nickel from ores that came from New Caledonia, supplied by the French company Le Nickel. The Erdington plant was owned by Le Nickel, as also was a plant at Kirkintilloch where the imported raw material was 'Bessemerised' – that is melted, and put into a Bessemer converter, through which air was blown to oxidise the most reactive impurities, leaving nickel, copper, iron and any noble metals in the melt. None of these plants survived the decline in demand after the war. Vivian's, who also processed

German-silver and other copper alloys at a rolling mill in Birmingham at Icknield Port Road, not far from Wiggin's works, carried on their copper interests. The Vivian business eventually became part of ICI Metals Division.

For the UK demand, the refinery at Clydach, using the carbonyl process on nickel matte imported from Canada, continued to be the biggest supplier of nickel, but other than International Nickel, the only other business to acquire a licence to use the Mond carbonyl process was the Russian company, Norilsk Nickel. Consequently, most of the nickel now produced is refined electrolytically.

The nickel-plating process operates by converting nickel ions in solution to metallic nickel on the surface of the conducting object being plated, however discharge of nickel ions is not the only process. The discharge of hydrogen ions to form hydrogen bubbles takes 3-7% of the current.

Under normal conditions the efficiency of nickel dissolution at the anode is 100% but, if the solution does not contain activating chloride ions, passivation can result especially at high anode current densities.

The simplest anodes are made by hanging nickel bars from hooks on an anode rail. The nickel anode is encased in a cloth anode bag to prevent the small quantity of insoluble residues from entering the solution which would cause roughness of the deposit.

Electroplated nickel coatings are widely used to improve the use, value and appeal of consumer goods. Decorative nickel coatings are smooth and mirror-like; when used as a base layer under chromium coatings they provide excellent corrosion resistance for consumer articles. Most nickel coatings are in the consumer area as they greatly improve visual appeal and add value.

Nickel coatings for engineering applications are usually smooth and matte and improve corrosion resistance. These coatings also improve the wear resistance of engineering components, often as a base layer underneath a thin chromium top layer to achieve excellent utility for the equipment which uses them.

In the 1960s the way that platers used nickel anodes was revolutionised using titanium anode baskets. The baskets are filled with small pieces of nickel which dissolve and flow down the basket. This development greatly increased the use of free-flowing nickel pellets made via the carbonyl process and INCO developed active forms of nickel anodes and forms which more easily pass down the basket, both by electrolytic refining and by the carbonyl process. Much of the development and testing of these new forms took place at the Wiggin Street Laboratories.

Another major development of the 1960s was the discovery and development of nickel sulphamate as an electrolyte for nickel deposition, especially for engineering coatings and for electroforming. This was first developed by Dick Kendrick at Nottingham University.[11] He subsequently joined the laboratory at Wiggin Street to develop and commercially exploit his process.

Electroforming is the use of electrolytic processes to manufacture tools and products which were often impossible to make by traditional methods of fabrication. This involves progressively electrodepositing metal onto a conducting mandrel to build up its thickness. At the end of the process the mandrel can be separated to leave a free-standing part.

The process can copy fine surface detail with great accuracy. An example of this is the production of electroformed nickel press plates for making audio records and digital CDs.

A good example, that will be readily understood, is the sub-masters for vinyl records. As is well-known the acoustic recording is made by amplifying the sound, causing a needle to vibrate while making a spiral track on a rotating wax disc. A very thin conducting layer applied to the wax allows it to become the cathode in an electrolytic cell in which the anode is nickel; the transfer of nickel to the surface of the wax master creates a negative image of the recording track, from which multiple copies of the original could be made by pressing a slightly heated thermo-plastic onto it. The similar copying method can be used employing laser engraved photo-resist plates to mass produce video and digital music discs.

The replication of holograms using this method exemplifies the extreme precision that can be achieved. The holographic token on bank notes and credit cards is made in this way.

The tools used in the plastics industry are often made of electroformed nickel because it allows the replication of complex surface finishes such as wood grain or leather patterns on the surface of the mould without the need for machining.

Many parts can be produced from a single mandrel with great dimensional accuracy and repeatability such as the small components shown in Figure 6.4.

These components made by the electrochemistry group at Wiggin Street include a hexagonal mirror, an array of which forms the basis of a Cassegrain reflecting telescope flown in space: a half ellipse used to collect scattered light off a surface and several small cones used for cooling CCD (charge-coupled device) image sensors in space applications. The figure also shows a cover for a waveguide. The mandrels for each of these parts can be used to make many copies of them.

Figure 6.4 *Some examples of small components produced by electroforming.*

The role of the Plating Section was to assist International Nickel's customers to solve their problems in using the company's products. As well as helping to develop the new anode materials that came to be used in higher speeds of electroplating and electroforming, the section pioneered new applications and new processes in support of INCO's customers.

One of the major developments at the Birmingham laboratory was high speed electroforming for the manufacture of thin nickel foil. Increasingly large machines were constructed to make nickel foil over the thickness range 4 to 100 microns. It involved plating onto a partly immersed, slowly rotating titanium drum cathode and stripping the coating as it emerged from the concentrated nickel sulphamate electrolyte. Initially machines were capable of 100mm wide foil but finally several machines capable of producing 500mm wide 4-micron thick foil at a rate of 100 metres/hour were constructed (Figure 6.5). The machines were designed by Peter Crouch and John Whittle. This is now the method that is used to make nickel foil.

Figure 6.5 *One of the nickel foil machines developed at Wiggin Street. (Courtesy Bob Giles)*

By plating onto a roll with a pattern of non-conducting dots on it, a strip with a regular array of holes in it can be produced – see Figure 15.1.

Another larger engineering development carried out at Wiggin Street was the manufacture of erosion shields to cover the leading edge of carbon-fibre, helicopter rotors. These are needed to compensate for the poor erosion resistance of epoxy-bonded carbon fibre components. In the production of the erosion shields the thickness distribution must be controlled to meet the design requirements. This is done by having the anode material, pellets of nickel, in titanium anode baskets that are placed in a controlled distribution around the cathode. To achieve the high mechanical strength and hardness required for these erosion shields, electrolytes were developed based on nickel sulphamate ($Ni(SO_3NH_2)_2$), with additions of cobalt sulphamate.

The successful development of these erosion shields for Westland Helicopters resulted in them being used on the rotor blades for both the Lynx helicopter and the EH101, which became the RN Merlin helicopter, and resulted in a manufacturing cell being set up at Doncasters Bramah (a production plant in Sheffield originally acquired by INCO). This has been in operation since 1989.

To address another emerging market, a process was developed to convert the surface of the nickel foil to a black oxide, which, it was found, absorbed more than 95% of the energy from light incident upon it. A product named Maxorb was marketed and became a preferred material in the manufacture of thermal solar collectors.

Of the energy consumed in much of Europe, about 40% is used to heat water, either for direct use or for space heating. There is a general, but erroneous, belief where this is the case that solar energy is not viable, because the air is generally too cold. Certainly, when the need for heat is greatest, the sun shines for only eight to twelve hours a day, but the light available during those hours is a useful source of energy. Its benefit as a source of heat has no relationship to the air temperature. If every house had a suitably positioned solar collector of the type shown in Figure 6.6 considerable inroads could be made into the demand for energy for space heating and hot water.

Figure 6.6 *One of the best thermal solar water heaters, produced in Northern Ireland by Thermomax.*

Inside each of the glass envelopes is a device in the form shown in cross section in Figure 6.7.

The selective surface, which could be black nickel oxide, captures the heat from the incident light and transfers it to a pipe made of copper. This contains a heat transfer medium; in the simplest case, this is water; the Thermomax design uses evaporation and condensation to transfer the heat to the manifold at the top of the array in Figure 6.6.

The glass envelopes are evacuated. Conduction and convection make no contribution to the performance of the device, so that any incident light, which is radiation, can contribute to providing heat to the water that circulates through the manifold and through a coil in a hot water tank.

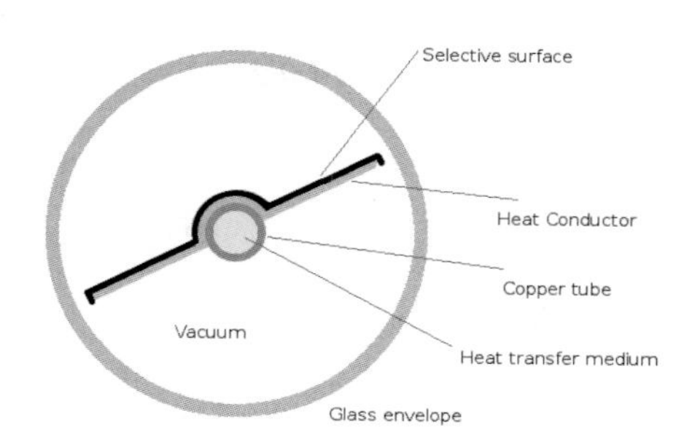

Figure 6.7 *Heating element in a thermal solar collector.*

When using the blackened nickel foil, it is attached to the heat conductor by adhesive. An alternative is to apply the nickel as a plated layer to the heat conductor before oxidising the nickel surface.

Many metals can be applied as coatings by electrolysis. Nickel is the one that is most connected with providing good service performance. Chromium is also used extensively as a plated layer, but if it does not have beneath it a layer of electroplated nickel, its performance is extremely poor. Cheap artefacts, such as door handles in quite expensive cars, are frequently found with the underlying plastic exposed because the chromium has peeled off. This does not happen when nickel is applied first.

7

Iron Comes of Age

The iron age started about 3000 years ago and iron has long been used to make tools and weapons, but it is only a little over 200 years since metal, in its various forms, became a serious structural material. Cast iron was used to make cannon guns from the middle of the fifteenth century, but it was poor stuff; the guns were about as dangerous to the firer as the fired upon. Wrought iron was used to make small arms and for any application where it was necessary to support loads. There were several early methods of producing a steel-like material, composed, like the cast iron then available, of just iron and carbon, with whatever adventitious ingredients came from the source materials.

Steel would then have been just a low carbon iron, while cast iron had significantly higher carbon levels. Carbon content would have been higher in the iron produced by smelting with coke after Abraham Darby (1678–1717) made this the common method. The product of the early coke blast furnaces was not suitable for engineering applications. The coke introduced not only more carbon, but also sulphur, which is damaging to the properties of iron. Poor sources of iron ore also introduced significant levels of phosphorus, also deleterious to strength. The sorry state of the famous Iron Bridge is testimony to the inadequacies of Darby's material as a structural material. Cast iron was, however, important in the early phase of the Industrial Revolution, but not until John (Iron Mad) Wilkinson (1728–1808) took advantage of a more important development to machine cast iron.

Benjamin Huntsman (1704–1776) developed a superior method of making steel, by building a furnace that could melt even pure iron. Cast iron from Darby's blast furnaces would have been molten above about 1200°C. To melt carbon-free iron, a temperature approaching 1600°C is required. Huntsman achieved this by placing bars of low-carbon wrought iron, and a form of iron that contained significant levels of carbon, together in a silica crucible and heating it over a coke fire. The heat was constrained to rise through a narrow gap around the crucible, driven by the pull achieved by building a tall chimney over the furnace. This, the crucible steel process, enabled controlled levels of uniformly distributed carbon to be achieved in the final mixture.

The first result of this development was the creation of steels that were excellent tools. Matthew Boulton (1728–1809) was perhaps the first Englishman to recognise the advantage of using Huntsman's materials for punches used to make fancy buttons and medallions and it was he that probably introduced John Wilkinson to the use of Huntsman's steel for the machining of cast iron. It was used initially to bore cylinders for Watt's steam engines, but subsequently to plane smooth surfaces on cast iron frames. These were important to Manchester's claims relating to the Industrial Revolution. Water power was used at first to drive several of the textile machines that were introduced in the western flank of the Pennine Range. A smooth, well-lubricated metal surface enabled more machines to be operated using this limited source of power.

Cast iron is not good for withstanding tensile loads, so frames to support machinery had to be designed to act in compression only. One such frame, almost certainly the one produced in the greatest numbers, was found in all but the poorest homes, and can still be seen in many cafés. It is probably the first flat-pack item for the home that was ever produced. An example is shown in Figure 7.1. It is a clever design. Unlike most modern flat-packs it needed no instructions on how to put it together. The treadle frame fits both ways.

However, Huntsman's process had a much wider significance. The crucible held about 40 to 50 pounds of liquid metal. It was easy with this configuration to make additions to the liquid, so for the first time it became possible to make alloys of controlled composition.

Figure 7.1 *A Singer sewing machine table and treadle.*

No doubt many attempts had been made over the centuries to add the metals of antiquity, copper, lead, tin and zinc to iron, all these metals being nobler than the iron. However, all have a damaging effect on the strength of the metal, either because they do not dissolve in the iron or they form brittle compounds. Like other more noble metals, nickel can be added to molten iron without loss to oxidation, and, unlike the others commonly available, it mixes freely at an atomic level with iron. Once it was available in quantity it is obvious that its use would be explored.

The first efforts at adding nickel to steels were early in the nineteenth century. Barraclough[1] suggests that a Swiss man called Fischer was the first to make alloys of iron and he had his London agent, John Martineau, take out a British patent in 1825 for 'meteor steel' that contained a tiny proportion of nickel. This claimed to reproduce the properties of the famous damascene steel, which probably means that it had a streaky appearance like the knife in Figure 7.2.

The better-known attempts to produce alloys of iron were attributed to Michael Faraday (1791–1867), who, with his co-worker J. Stodart[2] made alloys of steel with several metals. (The melting was done in crucibles in Sheffield at the works of one of the oldest Sheffield steel companies, Sandersons.) The nickel alloys had variable properties; the experiments

Figure 7.2 *A good quality chef's knife. The layered structure, produced by repeated folding and hammering of two different materials is typical of how knives were produced, combining a hard, cutting material with a soft, but strong one.*

were not well controlled. At that time the relevant information was that iron (or rather steel) and nickel could be mixed in any proportion.

These early attempts did not lead immediately to the development of an alloy-steel industry. Although Huntsman's method had been developed around 1742, it was not until 1859 that the first commercially successful alloy was made. Robert Mushet (1811–1891) added tungsten to the melt; he produced a 'self-hard' steel that became the basis for the tool steel industry. The tungsten combines with carbon in the iron, making a very hard carbide. (In this case the process did not involve metallic tungsten being added to the melt; the additive was its ore, wolframite $(Fe,Mn)WO_4$. The oxide was reduced in situ by the carbon contained in the steel).[1]

Percy reports[3], in the first significant book on Metallurgy in the English language, that his associate, a Mr Richardson, made alloys from iron wire and nickel containing 1, 5, 20 and 50% of nickel. The nickel, obtained from Evans and Askin, was said to be essentially pure but the iron wire is not commented upon. The samples were small buttons of between 300 and 500 grains weight (about a half to one ounce) and were evaluated by breaking the cast button; this was about as far as metallurgical science had progressed by 1864.

The first British patent for a nickel steel was awarded to Alex Parkes in 1870, but there is evidence of the earlier use of nickel in steel.[1] However the first publication showing details of the mechanical properties was not until Riley gave a paper on the subject to the Iron and Steel Institute in 1889.[4] Thereafter there was an explosion of activity that reached a peak in the 1914–1918 war.

By the time that James Riley (1840–1910) pointed out the benefits of nickel in steel, bulk steelmaking processes were in regular use. Bessemer's bottom blown converter and the open-hearth furnace developed in Birmingham by Charles Siemens (1823–1883), but first used to make steel by Emile Martin (1824–1915) in France, could now produce steel in tonnage quantities. These processes had already enabled the rapid expansion of the railway system, and in 1890 the first major civil engineering project to use steel, the Forth Railway Bridge, was opened. This used 54,000 tons of open hearth steel. The simple low carbon steel that was used was not a high strength material, as can be seen from the view in Figure 7.3a. The amount of cross bracing required to support the loads it had to carry is evident; as also is the extensive surface area that has resulted in the cliché 'painting the Forth Bridge' becoming the byword for the problem of maintenance. Ordinary carbon steel is not the best of materials for carrying heavy loads or for resisting decay. However, what is still the second longest cantilever bridge in existence has survived because it has been looked after.

The second, and longest, cantilevered bridge has fared less well. The Quebec Bridge is a Canadian National monument (Figure 7.3b). It was completed in 1917, at the third attempt, having collapsed during construction on two previous occasions with significant loss of life. To

(a)

(b)

Figure 7.3 *a) Showing the massive steel structure of the Forth railway bridge. (Photograph by Simon Johnston, CC BY-SA 2.0, https://commons.wikimedia.org/w/index.php?curid=14047927). b) The Quebec Bridge (Photograph by Sébastien Savard CC BY-SA 2.5, https://commons.wikimedia.org/w/index.php?curid=836550).*

finally complete the structure, 16,300 tons of nickel-containing high strength steel had to be used for the most highly stressed components.[5] It cannot now be used for carrying traffic because it is in a parlous state. It is a matter of dispute between the Quebec and National Governments about who will pay for its repair. The Quebecois clearly did not learn the lessons from the Scots about the need for maintenance.

It was the arms race at the end of the nineteenth century that drove the demand for nickel. The picture in Figure 7.4 is from a book describing its business, published by the Mond Nickel Company in 1918.[6] It celebrates the contribution of nickel in making armour plate resistant to armour piercing shells. There are five shells, all reduced to fragments and five pock marks on the gun turret corresponding to the attempts made by the five shells to breach the armour.

Figure 7.4 *Nickel steel armour plate, illustrating its shell stopping capability.*

The steel is proudly advertised as being Hadfield's patent Era steel. It was a cast steel. However, as with all developments in armaments, this is only half the story. In a lecture given to the Birmingham University Metallurgical Society in 1923,[7] Sir Robert Hadfield (1858–1940) himself showed a picture of the other half. He was equally proud of both sides of the matter. The picture is shown in Figure 7.5.

As with all examples of marketing, the message offered by these pictures needs to be taken with a pinch of salt. The plate pierced by the shell in a standard test would have been 12" thick, but it was a common grade of nickel free steel that had been treated to harden only its surface. In the armour plate trial, the steel was a through-hardened, nickel steel and the shell materials were not.

Nickel-steel armour plate was also used to clad battleships. The land ships that were developed under the auspices of the Admiralty also used this material. These were the first military tanks, their development encouraged by Winston Churchill, then the first Lord of the Admiralty. They were designed and built initially in Lincoln, which was then a centre of the production of agricultural machinery. However, most of the tanks used in WW1 were built in the Black Country.

Figure 7.5 *Robert Hadfield's picture of a shell made from a Hadfield steel that has passed right through armour plate intact.*

In the same period, the automobile had started to become a common sight, after Ford started to produce the Model T at Trafford Park in Manchester. Stronger steels were needed for some components and Henry Ford made a great deal of the use of vanadium steels to reduce the weight of the heavy components, such as crank-shafts. Where heat resistance was also necessary, for instance in engine valves, or to give better corrosion resistance, nickel steels have also been part of the development of the motor industry.

Since nickel is more noble than iron, it can survive in all the steel-making processes. The huge amount of nickel containing scrap that became available after the end of hostilities could therefore be used without significant loss of nickel, so that nickel bearing steels entered many load-bearing, structural and corrosion resistant applications. The nickel content of the steel scrap could be used and augmented in the production of stainless steels. (For other alloying additions that are less noble than iron, ferro-alloys were developed. These are generally added in the ladle following steel-making).

The benefits conferred by having nickel in the steel were improved strength and greater resistance to fracture. Iron based metals are susceptible to brittle fracture at low temperatures as are other body-centred cubic (bcc) metals.

Although metals are all crystalline in their common forms, few things made of metal are single crystals. It takes special care in processing to create a metal single crystal, as will be seen later. Common engineering metals are composed of many crystals, usually referred to as grains. Their structures are like those shown in Figure 7.6 where the individual crystals are revealed by polishing and acid etching the metal.

Figure 7.6 *Showing the individual crystals in a metal sheet.*

(The material in the figure is secondary-recrystallised silicon iron (transformer steel), in which the grains are as seen by eye without magnification, but similar structures with any size of crystal might be seen in annealed single-phase metals, usually under a microscope. The grain contrast is because the etchant attacks differently oriented crystals at a rate that depends on the orientation of the exposed surface.)

There are several different ways of hardening metals, which are described in section 2 of the Appendix, but there is only one way to make these metals both stronger and more resistant to brittle fracture; that is to refine the grain size. The effect of grain size is illustrated in Figure 7.7.

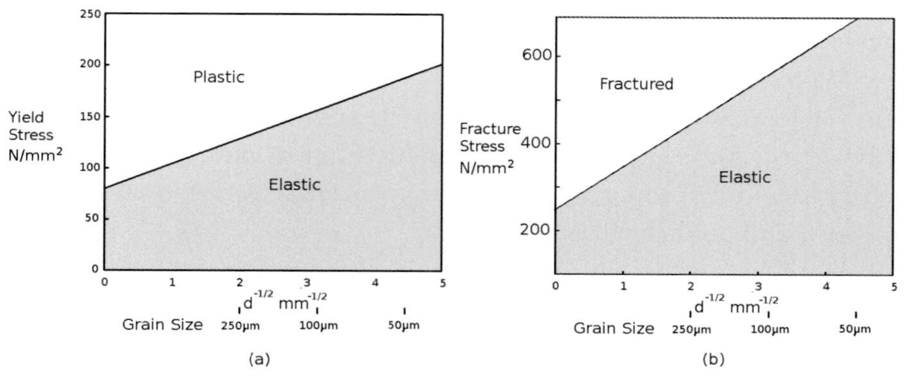

Figure 7.7 *a) The variation of the yield stress with grain size at room temperature for a mild steel. b) The fracture stress as a function of grain size for a mild steel at -77°C.*

The behaviour represented in Figure 7.7 shows that both the yield and fracture strength vary as the inverse square root of the grain size. The equations linking the properties to grain size are known as the Petch equations,[8] having been defined by Norman Petch (1917–1992). Although the two sets of data are not from the same steel, they do properly show that the strength is higher at the lower temperature. It is because the bcc metals have this strong temperature dependence that they become brittle. The higher strength cannot be used if the consequence is that failure can occur catastrophically.

There are several ways in which metals can be processed to give smaller grain sizes. Most of these have been developed in the second half of the twentieth century after the behaviour illustrated here had been understood. The significance of nickel in enabling the development of the properties illustrated in armour plate and missiles and in structural steels, was that they arose as a simple consequence of alloying without understanding how the improvements were caused. That understanding followed quickly from the recognition by Adolf Martens (1850–1914) of the structure of a quenched carbon steel, as seen under a microscope, which has, ever since, been called martensite. Nickel makes it easier to form martensite, which, when tempered, yields a very fine ferrite structure, interspersed with a fine carbide distribution. A fuller explanation is given in Appendix 2.6.

The benefits conferred by nickel in steel are also found in cast irons, of which there are several types. Nickel can be found in all the several varieties of cast iron, in all cases conferring improved corrosion resistance, improved strength and improved resistance to fracture.

With carbon levels below 0.8% the material could be cast but would be referred to as a steel. Cast irons usually have carbon contents between 2 and 4.5% and they can develop a variety of different structures and properties, depending on what other elements are present and on thermal processing.

Until late in the nineteenth century, the way the cast iron turned out was determined by selecting the source materials and keeping to the process that gave the desired results. But adventitious ingredients could result in the material containing either iron carbide (cementite – Fe_3C) or graphite. The cementite produced a hard material, while with graphite it would be soft, and if the graphite occurred in flake-form, it fractures easily along the planes of weakness in the graphite.

It was Thomas Turner (1861–1951), the first Professor of Metallurgy at Birmingham University, who discovered how to control this behaviour. He found that adding silicon to the iron resulted in the carbon being ejected from solution to precipitate as graphite. Since silica is found in many ore bodies, the presence of silicon in castings is not surprising. Silicon is similar in many ways to carbon; it is in the same column of the periodic table. With silicon above about 2.5% the carbon is in the form of graphite; at lower silicon contents the carbon is in the form of cementite. The two forms of cast iron can be detected from their fracture surfaces. The carbide containing form shows a white fracture surface, hence it is called white cast iron; the graphitic iron shows a dark surface and is referred to as a grey cast iron. (Turner described the finger test to differentiate the two.[9] A finger rubbed on a white iron would be clean; on a grey iron it would be blackened, as if by pencil lead, a mixture of graphite and clay.)

Most types of cast iron are unable to withstand tensile forces. Grey cast iron is especially weak in tension when the graphite flakes lie as lamellae in the iron matrix. They produce planes of weakness. However grey cast iron can be made more ductile if small amounts of magnesium or some rare-earth metals are added to the material. These influence the growth of the graphite plates during solidification through an effect on the surface tension between the liquid iron and the graphite. The result is that the graphite forms small spherical masses, which are much less damaging to the mechanical properties and create the form of iron called, unsurprisingly, ductile iron.[10]

Whatever the type of cast iron, nickel increases its strength and corrosion resistance. Other than the ductile irons, the applications are either where loads are compressive or where surface abrasion, or lack of abrasion, is required.

Grey cast iron with graphite exposed at its surface is used for machine beds and other surfaces on which there is movement between components. The iron can be easily machined to give a smooth flat surface, as John Wilkinson recognised, enabling good lubrication and energy saving in machinery. This form of iron would have been the predominant type until the twentieth century. Apart from the machine bases that Wilkinson would have supplied, the principal uses were in low performance applications, in competition with pottery and even wood. (The Iron Bridge is essentially based on a wooden bridge design). In more recent times a major application of the simple grey irons has been in cylinder blocks for internal combustion engines.

The form of grey cast iron in which the graphite is nodular, ductile iron, can be used for more demanding applications, such as gears and camshafts for cars.

White cast irons are used for highly abrasive applications, either to resist or to cause abrasion. For the most demanding applications, nickel containing irons are used. A composition containing 4.5% nickel, about 3% of carbon and 2% of chromium is called Ni-hard cast iron and is used in many industries, where materials are to be reduced to powders during processing.

Figure 7.8 *A white-iron slurry-pump case used to carry abrasive particles. (Made by the Taylor Group)*

A common application is for balls used in ball-grinding, which is used to comminute the material in wheat mills, cement mills and a host of other industries. In some cases, the particulates are mixed with a liquid carrier and pumped using a slurry pump (Figure 7.8). The most common application of a Ni-hard component being used to work with an abrasive slurry is the blade in a cement mixer.

The publications of the Nickel Institute[11/12] provide extensive information on the properties and applications for ferrous materials containing nickel.

The effects that nickel have in determining the properties of ferrous materials are discussed further in section two of the Appendix.

8

Metrology

a) Introduction

Measurement is among the most fundamental requirements of science and commerce. We measure how far and fast we travel in terms of length and time, what we buy in terms of volume, weight, or mass. All other units of measurement, except temperature, are related to these three. Standardisation of the units used began in France around 1890, and has evolved to the modern Metre, Kilogram, Second (MKS) system of the Système Internationale (SI). Local units of weights and measures evolved everywhere, but for international trade, what you pay for and what you get depends on having the same units in which to trade. Measurement is even more important in science. Without standards, communication of results among scientists would often be meaningless.

To ensure that the units used are consistent everywhere, each device used to measure length, weight or time has to be calibrated against a standard of that unit. The absolute standard is held by the Système Internationale and each country has a copy of this, calibrated against the one in France. Cheating on weights and measures is one of the most basic of frauds and developed countries generally have consumer protection organisations that have the responsibility of ensuring that the devices used by retailers are regularly checked.

In principle, a standard metre could be made of anything that is stable and durable. A standard of weight only needs to be resistant to corrosion; an iron weight initially rusting and making it heavier and then the rust peeling off to make it lighter would not do. So, what has nickel got to offer to the field of metrology? The answer is surprisingly extensive. Several examples will be related. The first one touches on two of the standard units, the metre and the second. Weight will feature in the derived units, but here the use of nickel is easily understood. It offers corrosion resistance, either through being plated onto a cheap iron standard weight or having the weight fabricated in a stainless material containing nickel.

b) Length

In 1920, the Nobel prize for Physics was awarded to Charles Edouard Guillaume (1861–1938) for his discovery of some anomalies in nickel-steel alloys. The next year the prize was awarded to Albert Einstein and in 1922 it was Niels Bohr's turn. On first sight, the name Guillaume sits ill among the great names of Physics, but the full citation 'in recognition of the service he has rendered to precision measurements in Physics by his discovery of the anomalies in nickel-steel alloys' makes it clear that the awarding committee was celebrating the importance of metrology in science. All science is based on observation and measurement, and as head of the International Bureau of Weights and Measures in Sèvres, France, metrology was Guillaume's business.

Guillaume's discovery was made in 1896, twenty-four years before the prize was awarded. In his acceptance speech,[1] Guillaume recorded that the first standard metre was made of a platinum-iridium alloy, which was very expensive. A cheaper substitute was needed. He decided that nickel offered the best option to study. He went on to give a detailed account of the researches, carried out by himself and others, that settled on a composition of 36% nickel and 64% iron (neglecting minor impurities that appear to be unimportant in this context). This has a very low coefficient of thermal expansion (1×10^{-6} compared to 10×10^{-6} for platinum), which makes it possible to use a standard over a wider range of temperatures. Guillaume named it Invar.

Until a few years before Guillaume's work, it had been assumed that a law of mixtures applied to alloys, although from about 1870 it was realised that iron exhibited some strange behaviour.[2] Nevertheless what Guillaume found was unexpected.

If steel or cast iron is force-cooled to a temperature a little below 723°C (see Figure 7.11) and held, it will warm up again. This phenomenon, called recalescence, occurs because latent heat is released until the material is fully transformed from austenite to ferrite.

A full explanation for this had to wait until X-ray diffraction became available to determine the crystal structures of the phases. It was still being studied in 1937, when Owen and co-workers used a high temperature X-ray camera to follow the expansion of a range of nickel-iron alloys. (Owen thanked Leonard Pfeil, Mond Nickel's Assistant Director of Research for providing his nickel). Owen's results are shown in Figure 8.1. The important result shown here is that the lattice parameter of the austenite containing about 36% nickel scarcely changes between 15 and 200°C.

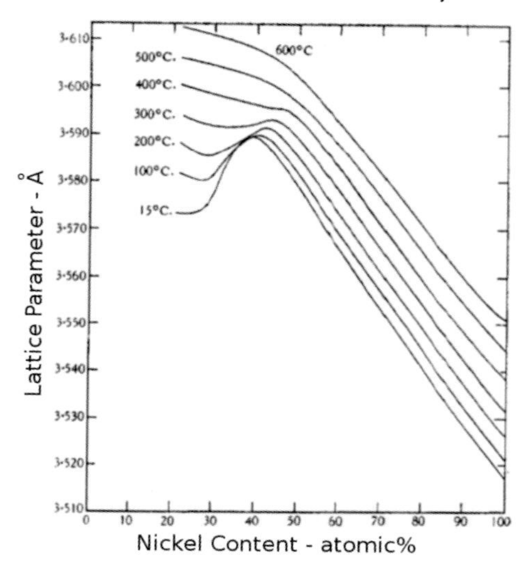

Figure 8.1 *The change of lattice parameter with temperature in austenite, as a function of the nickel content.*[3]

The use of X-rays allowed thermal expansion to be measured on individual components of the material. Guillaume could only use a dilatometer, which averaged the variation over the two phases present. However, by the time that Guillaume received his Nobel prize, the phase composition of iron alloys was better understood, mainly from the work of Chevenard (1888–1960).[4]

In his acceptance speech for the award of the prize Guillaume gave several examples of the uses that became possible from his discoveries. As the son of a clockmaker, it is no surprise that he was aware of the benefits in timekeeping, nor is it surprising that clocks were once again a driver of innovation. Huntsman's steel-making process was the product of his search – he was a clockmaker – for a better material for his watch-springs. The available steel at the time was prone to frequent breakages. His steel was subsequently used in Harrison's carriage clock, which was the solution to the longitude problem. Clocks were the first precision engineered products and many of the mechanisms of early industrialisation owe much to their designs.

Both in pendulum clocks and pocket watches, the timing mechanism depends on the length of a piece of metal. In the pendulum clock, it is the length of the pendulum itself. The use of Invar obviated the need to construct compound pendulums that were designed to compensate for changes in length due to changes in temperature. Similarly, in travelling clocks and pocket watches the length of the hair spring determines the periodicity of the motion. Harrison[5] (1693–1776) designed a bimetal balance wheel to compensate for temperature variation, which again should not be needed in a watch with an Invar spring.

For most of history, the standard of time was astronomically based, but since the development of the caesium atomic clock in 1955 the standards have been referred to that. But for many years carriage clocks similar to that produced by Harrison were used to synchronise time in different places. This became important when timetables for rail travel were produced. Now most people have clocks that are synchronised by signals from a satellite.

Guillaume cites several examples in surveying and metrology where the use of Invar avoided the errors resulting from the use of more common metals, but he relates how, in about 1912, he realised that even better control was possible if an alloy that was similar in expansion coefficient to Invar but had a coefficient of elasticity that remained constant over a wider temperature range could be produced. He went on to develop Elinvar, essentially Invar but with 6% of the iron replaced by chromium. The chromium strengthens the material, which results in its Young's modulus being less variable than that of Invar. Invar is rather soft and susceptible to creep.

The periodicity of the watch spring is affected by its elasticity and the result of Guillaume's development was to enable the manufacture of accurate timepieces without the use of bimetallic balance wheels. The alloy also has lower magnetic susceptibility than Invar, which is useful in some applications.

Elinvar is also the ideal material from which to make tuning forks. The note generated is determined by the dimensions of the fork and by the elasticity of the material from which it is made. If the elasticity varies the note will also vary.

Another use of nickel-iron alloys was suggested by Guillaume, from the studies he carried out. Following Hopkinson (1849–1898),[2] he had first used measurements of the alloys' magnetic properties to define the range for more detailed study by dilatometry. His measurements of the temperature at which the alloys became magnetic on cooling from a high temperature are shown in Figure 8.2.

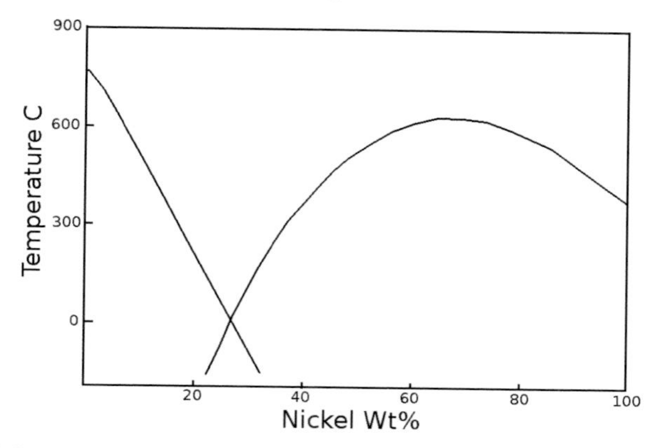

Figure 8.2 *Variation of the Curie temperature at which the alloy becomes magnetic on cooling in Fe-Ni alloys, as a function of nickel content. (After Guillaume)*

Guillaume suggested that this variation could be used for switching. An electrical contact sustained by a magnetic link would break when its magnetic strength fell below the level required to make contact.

The more common switching devices that use these materials, however, make use of bimetal strips. One side has a high coefficient of expansion, for instance copper, the other the lowest available, an alloy such as Invar. The strip changes its curvature as the temperature is raised, or cooled, acting to break a circuit of which the strip is a part. Every home probably has more than one: they are in electric kettles and in similar devices, in electric cookers and in central heating boilers.

One of the most significant applications for low-expansion alloys is in vacuum technology. It is common to need to introduce instrumentation or other metallic features into an evacuated glass envelope.

To do this a glass to metal seal is required. The thermal expansion of the metal passing through the glass has to be matched to that of the glass, which is very little.

The behaviour of nickel-iron alloys has been a matter for study by mathematical physicists ever since Guillaume discovered Invar. The explanation lies in the field of quantum physics, of which its first proponent, Niels Bohr famously said, 'if anybody says he can think about quantum physics without getting giddy, that only shows he has not understood the first thing about them.' Even those who claim to be expert in the subject have difficulty in giving a coherent explanation of what happens around the Invar composition. Some idea of the underlying principles is given in Section 1.3 of the Appendix.

c) Electrical Properties

Volta's development of a chemical source of electrical current opened a new field of study that attracted many scientists early in the nineteenth century. It was not until 1827 that Georg Ohm (1789–1831) published the law that bears his name, relating current (I) to voltage (V). In its current form this is expressed as:

$$V = I.R$$

where R is the resistance in the circuit. This is, naturally, measured in Ohms. V is in volts and I in amperes.

Now electrical circuits could be designed, and it became necessary to accurately measure these parameters. As technology progressed, the designs became more demanding and it became necessary to have standard resistors that enable the calibration of unknowns. This is done using the Wheatstone Bridge. This was originally conceived by Samuel Christie (1784–1865) but is named for Charles Wheatstone (1802–1875) because he first proposed its use for calibrating the resistance of unknown components. It was not until 1887 that Edward Weston (1850–1936) discovered that some alloys had a resistivity that was insensitive to temperature over a useful range around room temperature and developed two alloys that could be used as resistance standards. In the Wheatstone Bridge (Figure 8.3), there are three resistors made of one of these alloys, one of them variable, a fourth one is the unknown.

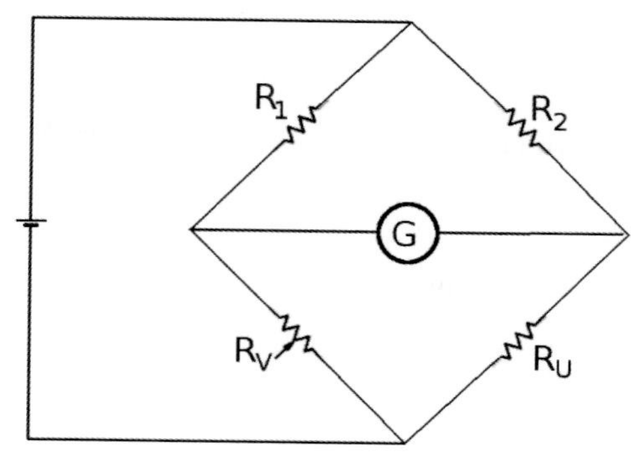

Figure 8.3 *A Wheatstone Bridge circuit. The variable resistor R_V is adjusted until the galvanometer (G) shows that no current is passing through it. The unknown resistance, R_U, is then given by $R_V/R_1 = R_U/R_2$.*

The resistance of the variable is linearly dependent on the length that forms part of the circuit. All of the standard resistors are made of one of Weston's alloys. These are Manganin (86% copper, 12% manganese and 2% nickel) and Constantan (55% copper and 45% nickel). Both feature nickel, but, as will be seen, constantan has the widest range of uses.

Copper and nickel are mutually soluble in all proportions. Copper is among the metals with the least electrical resistance (1.75×10^{-8} ohm metres), nickel somewhat higher (6.93×10^{-8}) but the mixture of 55% copper and 45% nickel has a much higher resistivity at 49×10^{-8}. This high resistivity makes the alloy suitable for many applications in addition to being used as a standard. It is widely used in the windings of electric motors and in rheostats, because of its high resistance, temperature invariability and especially the fact that the alloy is resistant to corrosion. Various trade names have been given to this and similar alloys. Ferry is the name used by Henry Wiggin & Co.

When a current is passed, the electrons migrate under the potential difference at a rate that depends on how smooth their path is. The conduction electrons are moving through a space occupied by atomically bound electrons and in the charge field of the positive ion core, so they are moving through a crowded space. If the space is a regular array, as it is for a single species of atoms, the steps along the way will all be similar. Evidently these steps are easier in copper than in nickel, but in Constantan the steps are less regular. The moving electron will meet different atomic species and different electron densities at irregular intervals, impeding its progress.

The application of heat causes the ions to vibrate more on their lattice sites and this accounts for the increasing resistivity with increasing temperature. Constantan has a lower dependence on temperature, probably because the effect of temperature is less significant when the impedance is already so high.

d) Stress and strain

In common parlance, stress, strain and pressure have similar meanings. Technically they are very different. Stress relates to the load applied to an element in a load bearing structure; strain is the distortion that results. Pressure is the application of the same stress in all three directions at right angles to each other.

To maintain awareness of the integrity of any critical structure it is usual to monitor the strain at vulnerable positions and this is done using strain gauges. The strain gauge is a resistor, attached using a flexible adhesive to the part that is being monitored. In the simple case, where the part is under tensile load and is being stretched, the strain gauge elongates in parallel, resulting in its resistivity changing. The change in resistivity is a simple consequence of the wire, or foil, of which the strain gauge is made, becoming longer and of smaller cross-section. Its resistivity is measured using a Wheatstone Bridge. Both the reduction in cross-sectional area and the increase in length raise the resistance, so that it rises as the square of the strain.

A high basic resistivity is required for good sensitivity and, furthermore critical structures, like buildings and nuclear power plant pressure vessels do not generally exist in a controlled temperature environment; it is therefore important that the strain gauge material is not temperature sensitive. Once again Constantan is a common material of choice for this application.

e) Temperature

The accurate measurement of temperature was not possible until Gabriel Fahrenheit (1686–1736) produced the first mercury-in-glass thermometer in 1714. This method of measurement has a limited range, so to control his pottery kilns Josiah Wedgwood (1730–1795) devised a method, using pyramids made of different mineral combinations that slumped at different temperatures, to determine when a kiln reached the desired temperature. Accurate measurement up to the temperatures of industrial processes became possible after Thomas Seebeck (1770–1831) observed, in 1821, that if two dissimilar metals are joined together at different temperatures a voltage is produced.

The metal pair produces a voltage that is specific to that combination and changes with the temperature difference between the hot and the cold junctions, and it was discovered that the most useful combinations are different alloys of nickel.

In Figure 8.4 the variation of voltage with temperature is shown for several different alloy combinations. These are calibration curves, which allow the temperature to be determined from the measurement of the voltage produced when one end of the couple is inserted in a hot region, while the other is kept at a standard low temperature.

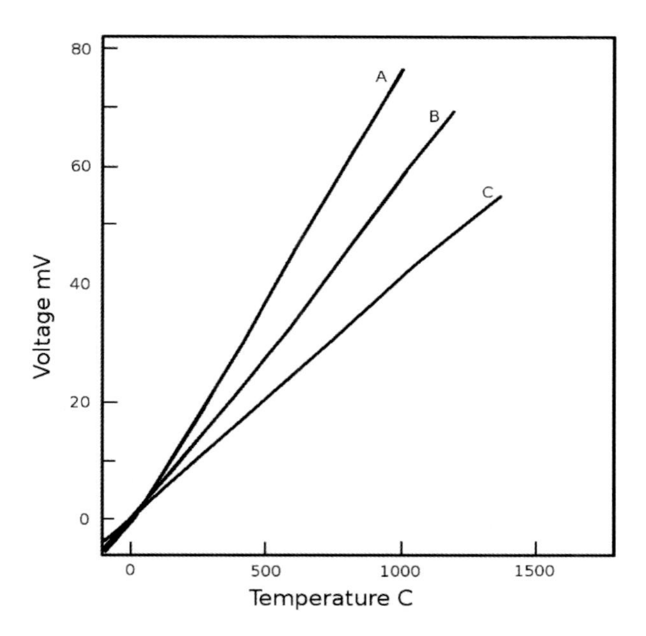

Figure 8.4 *Voltage versus temperature for three different thermocouples. A is for chromel-constantan, B for iron-constantan and C for chromel-alumel.*

Except for the use of iron wire in thermocouple B, the other alloys all contain nickel. Constantan is 55% copper, 45% nickel, chromel is 90% nickel with 10% chromium and Alumel is 95% nickel with 2% each of manganese and aluminium and 1% of silicon. The most widely used is Chromel-Alumel, which has the widest temperature range although not the highest sensitivity. The least common of the three shown is B. This points to one of the reasons that nickel features in the preferred materials. It has the best resistance to oxidation, except for the noble metals.

Although the use of a thermocouple may only involve intermittent exposure to the temperature that is being measured, iron would not be a good choice for use in an oxidising environment; it would certainly not last long in the applications in which the temperature measurement is continuous, for instance in controlling a heat-treatment furnace.

Following Seebeck's observation of the presence of a voltage in a couple, Peltier (1785–1845) observed that applying a current to a couple resulted in heat being generated at one of the junctions. This is a corollary of the Seebeck effect and is due to the same phenomenon. One junction is heated while the other is cooled. This makes it theoretically possible to make a heat pump simply by passing a current through a circuit consisting of two dissimilar metals. Refrigeration and air conditioning are possible uses of this principle. However, the conventional condensation – evaporation and absorption – desorption circuits are more efficient, and the use of Peltier heat pumps is limited to special situations.

A simple, but reasonable explanation of the Seebeck effect can be given in terms of the band theory of the valence electrons in metals. The energy level of the electrons will be different for the two metals. At the junction, the electrons in the side that has the higher energy will encounter an environment where they can lower their energy by moving to the other side, thus the lower energy side will have more electrons than required for electrical neutrality, the higher energy side fewer, hence a charge difference is established, which is the voltage difference. Raising the temperature of one junction raises the electron activity, pushing more electrons across the interface.

The Peltier effect, when a current is applied, now requires the electrons to circulate. They fall over from the high-energy side with the release of energy – giving a temperature rise, but to get back, electrons must have their energy raised to pass from the low energy side to the higher energy state, requiring the input of energy from the environment, which becomes cooler.

There are now many other ways of measuring temperature, using solid-state devices, but the nickel alloys remain important in all areas of metrology.

9

Heating

In Pana, Illinois, Albert Marsh (1877–1944), an alumnus of Pana High School, is celebrated as the father of the electrical heating industry.[1] He, it is claimed, invented an alloy, which he called Chromel. It was composed of 80% nickel and 20% chromium. (The alloy referred to as Chromel in chapters 5 and 8 has the 90/10 composition.) He was granted US patent 811,859 and UK patent 2129 for the alloy in 1906.[2]

The patents are very widely drawn, claiming up to 50% chromium, or other elements that he calls the chromium group metals: molybdenum, tungsten and uranium, and up to 50% of either nickel or cobalt. He claimed that it mattered not whether nickel or cobalt was used but concentrated on nickel because it was cheaper. He also focussed on alloys with up to 20% chromium as being suitable for his aim. This was to make a better electrical resistance material for heating purposes.

Marsh's patents were challenged by Elwood Haynes (1857–1925), who claimed prior art. However, he decided to concentrate on cobalt alloys and went on to develop Stellite, an exceptionally hard alloy that mainly uses cobalt and another of Marsh's list of metals, namely tungsten.

It is unlikely that Marsh received much in royalties; from 1905, Henry Wiggin and Co had installed rolling mills and wire drawing equipment at the works in Wiggin Street and had begun to produce the range of nickel-based wires that were in demand following the developments made by Guillaume and others. 80/20 and 85/15 nickel/chromium alloys were included in the product portfolio and were marketed as a range of Brightray alloys. These alloys are reported in other places as having been invented at Wiggin Street; Brightray was a registered trade mark of the company. Wire became one of its staple products and a large part of the Wiggin Street site was given over to this activity for many years, as shown in Figure 9.1.

Figure 9.1 *Henry Wiggin & Co wire mill in 1935.*[3]

The virtue of the nickel-chromium alloys is that they resist oxidation up to a very high fraction of the melting temperature and have a high electrical resistance.

Passing a current through a wire generates energy given by:

$$W = I^2.R$$

where W is in watts, I in amperes and R in ohms. The heat generated finds many applications.

The 80/20 composition can be used continuously at temperatures up to 1150°C, which is rather higher than required for most domestic applications. In consequence, these are more commonly supplied with an alloy that is much cheaper, being as much as 42% iron, with 37% nickel and 18% chromium plus other minor additions. This alloy is rated to perform at temperatures up to 1050°C, more than enough for toasters, kettles and radiant fires.

The age of electricity really took off with the range of applications that these alloys developed and in 1927, Mond Nickel, by now having purchased Henry Wiggin & Co, took a rather different approach to encouraging the market for its materials: it founded a company to make electric furnaces. The Birmingham Electric Furnace Company was set up on Tyburn Road, the main road from the city centre to the north-east. It became the largest such company in the UK. Most of the buildings still stand and are shown in Figure 9.2. The office block, which can be seen on an aerial photograph taken in 1937,[4] has been demolished. It stood where the car park is now marked out.

Figure 9.2 *The buildings erected for the Birlec plant lying alongside Tyburn Road and the Birmingham to Fazeley canal as they remain today. (Picture from Google Earth, Map data 2018).*

Although the attraction of using resistance heating for industrial furnaces probably led to the Mond Nickel Company creating Birlec, as it became known, it was not long before it was realised that the access the business had to the market made it essential that it supply all types of furnace, whether they assisted in the sale of the nickel alloys or not. Earlier furnaces had used a fossil fuel or wood to provide the heat, which had been transferred to the material being heated by either or both radiation from the flame and convection from the combustion gases. The electrical source of power offers other possibilities.

The first advantage, which applies to resistance heating, is that the atmosphere in which the heating is carried out does not have to be composed of combustion products. Filling the space with an inert gas retains a contribution of convection, but radiation is the main source of heat. The radiation still works if the material is in a vacuum. An active gas can also be used, for instance, to carburise the material. (In this case the resistance of nickel alloys to carburisation is an advantage). With these advantages, electric resistance furnaces are mainly used for heat treatment of materials in the solid state.

Melting furnaces do not especially use nickel base alloys. They are either fired using fossil fuels or, if electric, fall into two main categories: arc furnaces and induction furnaces. Both of these electric furnaces can work in a gas atmosphere or in vacuo. They are of interest here, not because they are constructed using nickel alloys, but because their invention was an essential step in the development of nickel-based materials, as will be further discussed.

Figure 9.3 *The canal tunnel carrying the extension to the Birlec factory on Tyburn Road.*

Birlec was a very successful company, which needed to expand to meet wartime demand, including that from the nearby factory in Castle Bromwich, where most of the Spitfires were built. The plant could not expand on either side, constrained as it was by a canal and the main road (until the building of the M42 it was the only major route from the south-west of the country to the north-east; it passes through the middle of Birmingham). The first attempt to solve this problem was the building of a bay with a tunnel for the canal running under it (Figure 9.3). The building is still there. It is supported on 53 reinforced-concrete arches and is about 200 metres in length.

By the time this extension was built, Birlec had outgrown its relationship with the nickel industry and others had moved aggressively into the electrical engineering business. In 1954 it was sold to AEI (Associated Electrical Industries), which relocated the business to Aldridge, before succumbing in 1967 to a takeover from GEC, which soon closed the plant.

This did not, of course, diminish the importance of the nickel-chromium alloys to Henry Wiggin and International Nickel, nor their importance to virtually every aspect of modern life. Although they are hidden from sight in most circumstances, very little that we have come to rely on would have been developed without these alloys.

One of the most significant inventions, which should be more widely appreciated, is the thermionic valve. Invented by J.A. Fleming (1849–1945) – he of Fleming's left-hand rule – in 1904, this was the most important step in making wireless communication reliable. It was the beginning of controlled semi-conduction. The quartz crystal used in early radios was working, when it did, by rectifying the signal from the radio waves, but a device that only allowed current to pass in one direction in a controlled way was necessary for reliable reception.

There is a strong Birmingham connection to the development of radio. Heinrich Hertz (1857–1894) is credited with being the first to demonstrate the validity of Clerk Maxwell's prediction that radio waves existed; his priority is based on publishing first. This was an idea that was certain to be pursued after Maxwell explained Faraday's observations of the interaction of electricity and magnetism, and Oliver Lodge (1851–1940), the first principal of Birmingham University, carried out experiments at the same time as Hertz, which demonstrated the existence of these waves. Lodge was also the first to use the waves to transmit a signal and Marconi (1857–1894), who gets all the credit for early work on radio, was obliged to recognise Lodge's patents on the means of receiving the waves.[5]

Fleming's first diode consisted of a carbon cathode and a platinum anode in an evacuated glass tube. The cathode was directly heated to a temperature at which electrons are emitted. Application of a voltage between the cathode and the anode causes electrons to flow to the anode but reversing the voltage does not result in a current in the opposite direction. The materials that are good electron emitters, that is that have a low work function, do not generally make durable components, so a few years later H.J. Round (1881–1966) developed a better solution, using a separate heater to raise the emitter to the required temperature. He also independently developed the triode, which has a grid to which a voltage can be applied, lying between the cathode and the anode. This enables the triode to be used in amplifying the signal. It is shown in the schematic in Figure 9.4.

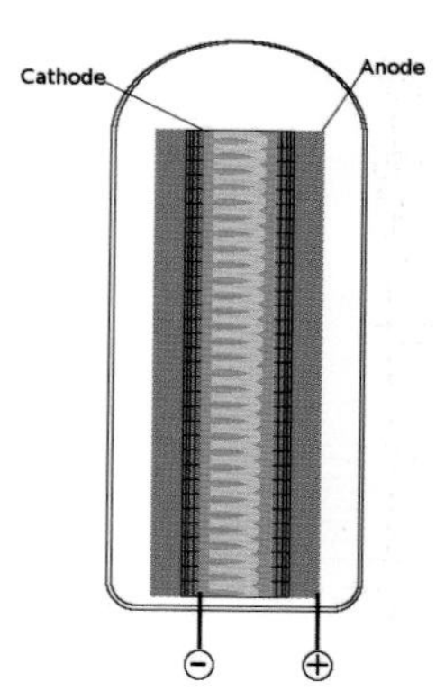

Figure 9.4 *Schematic of a section through the structure of a triode.*

The use of a heater made the device much more reliable. Without this modification, the most important means of communication and of entertainment from about 1930 to the mid-50s would have lacked an audience. In the early days, wireless was powered by lead acid batteries and it was a common sight even in the late 1940s to see people struggling with these heavy objects to take them to be recharged. The power was required to heat the coils in the radio valves.

Perhaps the greatest testimony to the importance and reliability of the indirectly heated thermionic valve was in introducing the digital age that had been made possible by the work of Alan Turing (1912–1954). It was he, who realised that everything could be represented by strings of zeros and ones; he was the father of digitisation and hence of computing. He further realised that to make a device that would work only in noughts and ones, he simply needed a collection of on-off switches. The first crude attempt to develop his code-breaking machine used switches, which took a very long time to go through all the combinations to decipher the German code. It was Thomas Flowers (1905–1998), a Post Office engineer, who realised that the most effective way to build a digital device was to use arrays of thermionic valves. He had already pioneered exchange switching apparatus for the Post Office before the need for code breaking at Bletchley Park. The machine built by Flowers, Colossus, was the first digital machine ever built and was a vital contributor to the Allied war effort. There were eventually ten Colossi working at Bletchley Park.[6] After completion of the Mark II prototype the remainder were made at a Post Office factory in Birmingham. The first model (Mark I) used 1700 valves, later models 2400.

The first computer to run a stored programme was made at Manchester University; a model of it, the Manchester Baby, is in the

Figure 9.5 *Model of the Manchester Baby, which was the first digital computer to store programs.*

Manchester Museum of Science and Industry and is shown in Figure 9.5. It also used thermionic valves, in this case pentodes.

It was not only thermionic valves that operated on the principle of the one-directional flow of electrons. The cathode-ray tube has an electron emitting cathode, heated by a nichrome element, which under the influence of a voltage accelerates the electrons towards a phosphor coated screen. The electron beam is moved by a magnetic field that is altered by the applied signal so that it creates an image on the screen that reveals the nature of the input signal as a function of time. The result is familiar to all in its use to reveal the heart rate of hospital patients and to indicate the return radar signals that show the whereabouts of neighbouring aircraft or boats.

More sophisticated cathode ray tubes were required for the creation of a reliable television service. Many people had a hand in developing television, but it was not until 1936 that the BBC introduced the first regular high definition television service. This used cathode ray tubes with a hot cathode, first made by Western Electric in the US.

The television images that were transmitted until this century were analogue in character, a single electron beam leaving an impression of light or dark depending on the intensity of the signal corresponding to the position on the screen. Today the transmission is digital, building further on the revolution begun by Turing, but requiring a very different technology. Thermionic valves made it possible to operate digitally, but it was the discovery of semi-conducting materials that eventually made digitisation ubiquitous. The key role played by the nickel-chromium alloys in making so many modern necessities available is now largely a matter of history. There do remain several small components in all our television sets that use the alloys as resistors and all our electrical heaters still depend on these alloys.

For heating applications, notably electric cookers, a common configuration is to enclose the heating wire in a tube. This is most often made of an alloy designated as Inconel 800. It is advertised as being resistant to oxidation and carburisation. There are several Inconel alloys that will be met later, all were developed at International Nickel's laboratories in the USA and were initially manufactured at its two metal plants in Huntington West Virginia and at Henry Wiggin and

Company. Several of these alloys are very close relations to the Brightrays; Inconel 800 is little different from the high iron version that has been mentioned.

In a recent book, entitled 'The Most Powerful Idea in the World', William Rosen advances the theory that industrialisation was a consequence of the creation of the patenting system, which turned ideas into property.[7] He may be right, but there is a good case for suggesting that in alloy development it is an idea that has gone mad. Very small differences in properties have resulted in the creation of many patents and a confusing array of choice for design engineers, who could almost always work around their problems with a much smaller range of options. Alloy developers are little different from alchemists, trying by a different route to make gold from base metal. The first to open a new avenue of study invariably generates several copyists, each of them striving to demonstrate originality. Perhaps Albert Marsh does deserve some credit for being the first to patent nickel chromium alloys, but everything that started from this base was then inevitable. This includes the whole family of stainless steels, which will be considered in the next chapter.

Chromium is a partner for nickel in many of the important alloys. It has many compatible characteristics.

In iron based alloys, it coexists with the nickel in solid solution, as it does in the alloys used in heaters. Its effect in resisting corrosion is to form a passivating oxide layer on the surface of the metal. The oxide Cr_2O_3 is very adherent, but if scratched or otherwise removed, it will reform, with the added benefit that the nickel-rich matrix is itself resistant to corrosion.

When chromium plating is used, the layer formed on plastic or steel has poor adhesion. Nickel used as an inter-layer between the artefact and the chromium gives good adhesion to the substrate, and the chromium layer and the nickel fit well together, making a much better product than if the nickel is not included.

Chromium has five d electrons and a single s electron and can form compounds with a valency up to six, using these electrons. (Hexavalent chromium is toxic and is the reason for environmental concerns about chromium plating). In its common oxide, it is trivalent.

Because of the variability of its valency, the atomic size can be variable also. In its usual body centred cubic form the atomic size is little different from that of nickel, which would suggest that the hardening effect of chromium in nickel should be slight. However, a hexagonal form of the metal has a considerably larger atomic size.[8] Since the number of near neighbours in the close packed hexagonal structure is the same as in the face centred cubic, it is probably at the larger size that the chromium goes into solution in nickel. In any event the addition of chromium to nickel has a stronger hardening effect than, for instance, copper. Copper atoms are about 2.5% bigger than nickel, the larger chromium variant is 11% bigger.

10
Stainless Alloys

The first nickel alloy to find engineering applications based on its corrosion resistance was composed of copper and nickel in the ratio of one of copper to two of nickel, plus small amounts of trace elements. Like the Chinese Pai t'ung, the alloy was developed because the roasted ore contained the ingredients in the proportion that the alloy inherited. The original intention was, indeed, to reproduce a material like Pai t'ung. Reducing the nickel-copper matte to metal with charcoal gave a master alloy to which only the addition of zinc was required to make nickel-silver. However, it was found that the nickel-copper alloy was itself of interest. It was brighter than nickel, very strong and resistant to acid attack and particularly resistant to corrosion in seawater. It was patented in 1906 in the name of, and given a name derived from, that of the president of International Nickel, Ambrose Monell, although the alloy was discovered by David Browne under the supervision of Robert Stanley. The alloy is still called Monel.[1]

The alloy has found very wide application. In marine applications, it is used for propeller shafts and propellers, in power plant it is used for furnace tubes, in architecture it has been used to roof many prestigious buildings and, more surprising, it is used by some guitar players as a string material, while the best trumpets and French horns use Monel for the valves.

The ability to vary the properties of alloys by heat treatment had been known for centuries for ferrous materials containing carbon.

Following the chance discovery by Alfred Wilm (1869–1937) that an aluminium-copper alloy grew stronger when left at room temperature after quenching into water from an elevated temperature, heat treatment of non-ferrous metals became a subject for study. The phenomenon, age-hardening, was found later by X-ray studies to result from the formation by diffusion of clusters of atoms in ordered arrays. These interfered with the deformation process to cause a significant increase in strength.

Wilm's discovery was followed up at the US Bureau of Standards by a team led by Paul Merica (1889–1957), who found that this type of behaviour was not limited to aluminium alloys. He subsequently joined International Nickel and led the research at the Bayonne laboratories for several years. When research was moved from Bayonne to a new facility in New York State, it was called the Paul D. Merica Laboratory.

Among the many material improvements that Merica is credited with, is the first age-hardening nickel alloy that uses aluminium and titanium as the hardening ingredient, Monel-K. This was the forerunner of the most important of alloys that became critical to the development of aerospace, as will be described later.

A relatively small tonnage of nickel now finds its way into copper-nickel alloys, other than those used in coinage. By far the largest market for nickel is now in alloys that use it in only modest proportions. About two thirds of the nickel produced is used to make stainless steel.

Monel and stainless steel both have applications that are principally cosmetic means of keeping the rain out of buildings. Several prestige buildings in New York, Washington and Philadelphia have Monel roofs.[1] Perhaps the best known of these is the Pentagon in Washington.

Two thirds of annual nickel production is about 1.2 million tons. The production of stainless steels in 2015 was about 41.5 million tons,[2] so that on average stainless steels contain less than 3% of nickel, which indicates that it is not the nickel that confers the property of resisting staining, it is chromium that is responsible.

The effect of an oxide coating inhibiting the corrosion of the underlying metal, called passivation, was first noted by Christian Schönbein (1799–1868) and reported in a letter[3] to Michael Faraday, dated April 22nd, 1836. Most oxides do not have a persistent passivating effect, for when they are penetrated the corrosion continues. This would be the case for a chromium-plated layer on iron, but when the chromium is also contained in the underlying metal, the passivating coating can reform when breached.

In principle, it is possible to add chromium to all the iron-based materials discussed in chapter 7 to make them stainless. The strong affinity of chromium for carbon requires that even stronger carbide-forming elements must be included if much of the chromium is not to be combined as a carbide in the higher carbon materials, but having enough chromium in solid solution does make the alloy stainless.

The first commercial alloy to incorporate significant levels of chromium was patented in Germany in 1911 by W. Borchers and P. Monnartz.[4] It was Monnartz who first noted that a minimum level of chromium, 10.5%, was needed to give the good corrosion resistance.[5] The patent was assigned to Krupp, whose researchers went on to patent alloys containing both chromium and nickel in 1912.[6] The first British patent was not until 1913; it was granted to Harry Brearley (1871–1948) for a steel containing 13% chromium, which he promoted for use in cutlery. A similar composition was subsequently patented in the US by Ellwood Haynes, leading to a dispute between him and Brearley, who also attempted to patent his alloy in the US in the same year, 1915.

The original Brearley and Haynes patents were for martensitic steels. Having contested the nickel chromium alloy patents of Albert Marsh, it is rather surprising that it took so long for Haynes to patent the cheaper alloys that also benefit from the passivating chromic oxide. Sadly, it is not surprising that Sheffield had not recognised this as an avenue that should be explored. As William Rosen pointed out, Huntsman did not patent his process but preferred to try to keep it secret.[7] Rosen evidently was not aware that this tended to be the Sheffield way. Dulieu[6] makes several references to Brearley being unaware of developments elsewhere; there was even evidence of a director of his company, Firth Brown, having visited Krupp in 1913 but not telling Brearley that he had seen samples of their alloy. Playing their cards close to the chest was always a characteristic of Sheffield steelmakers. A notable indication of this was the animosity that was shown to Henry Seebohm by the Sheffield steelmakers after he presented a paper to the Iron and Steel Institute in 1884, revealing the details of Huntsman's crucible steel process.[7] This was almost 30 years after Bessemer told the world about his process and a hundred and forty years after Huntsman's development.

Early stainless steels would not have had a low carbon content. It is, as described in chapter 7, the carbon that enables martensite to be formed on rapid cooling in most, but not all, steels. The first paper that promoted the use of nickel in steels was entitled 'Alloys of Nickel and Steel'.[8] At low carbon levels the steels with 12% or more chromium solidify as ferrite and generally remain ferritic at all temperatures unless very slowly cooled, when they will segregate into iron-rich and chromium-rich body centred cubic phases. Some ferritic stainless compositions contain a little nickel. This stabilises the iron-ferrite, but only at low nickel levels. Higher levels of nickel produce the face centred cubic phase, austenite. By varying the amount and proportions of chromium and nickel, it is now possible to produce stainless steels that are ferritic, austenitic or a combination of the two. The latter are called duplex stainless steels. These steels all find applications, but it is evident from the low percentage of nickel in the total output of stainless steel that the market is heavily biased towards the ferritic and low nickel varieties.

There is no doubt that the austenitic grades are generally superior; this is especially the case for applications that require superior corrosion resistance in acid conditions and at elevated temperatures. However, they are more expensive. At one time, it was common to find the rubric '18/8 stainless' on spoons and forks, but today the variety of steel is rarely mentioned. The practice, common in the fashion industry, of putting a designer's name on a cheap product made in the Far East and selling it at an inflated price now penetrates all areas of business.

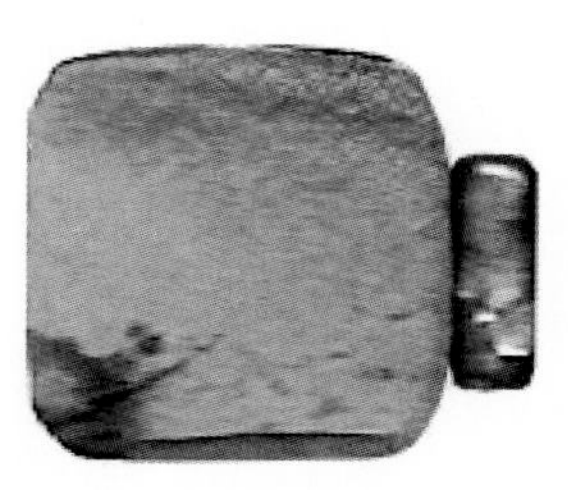

Figure 10.1 *Fracture surface of a tine from a garden fork. A small magnet is attached at the right. The dark stain shows the pre-existing forging defect.*

Unfortunately, this no longer means that you get what you pay for, as illustrated by

Figure 10.1. This shows the fracture surface of a tine from a garden fork. The first time the fork was used on a very warm day to lift earth, the tine broke off at a forging defect. The fork had the label of a highly-reputed manufacturer on it but turns out to be made from a most unsuitable material, martensitic stainless, a brittle material with no defect tolerance. This would not have happened with an austenitic steel. The catastrophic failure could have caused injury to the digger. An austenitic steel may have bent because of the defect but it would not have broken. The steel was almost certainly a cheap import and the fork may also have been produced in the Far East.

Stainless steel is now found everywhere. In the home, it is the material of kitchens. Cladding of ovens, washers and refrigerators, the sink, washer drum and dish-washer and micro-wave oven liners are all stainless steel.

Cooker hoods, as shown in Figure 10.2, are a common sight in modern kitchens, and like the knives, forks, spoons, pots and pans are commonly made of the ferritic types. More serious applications, such as boilers and fermentation vessels, would be better using austenitic grades.

***10.2** A modern gas cooker with hood, splashback, cooker plate and pots all in ferritic stainless.*

Modern kitchen ovens are clad in austenitic stainless. It is common to combine electric and microwave ovens side by side and it is then necessary to use the austenitic grade, because it is not ferromagnetic. The magnetron in the microwave oven operates in a strong magnetic field.

Chemical engineering plants, especially at elevated temperatures, are usually composed of austenitic steels; for the highest temperatures, as for instance in reformer tubes, high nickel alloys are needed to survive prolonged exposure.

In transport applications, the most stainless steel probably goes into exhaust ducts, which are made of ferritic steel, while tankers could be either ferritic or austenitic depending on the material to be carried.

Cutting tools for use in clean environments are usually stainless. This includes razor blades and medical instruments.

Perhaps the most obvious stainless-steel application is in the cladding of buildings. As a means of keeping the rain out, it has the advantage that it requires very little maintenance. There is, however, a downside to its brightness, experienced by those who suffer from the reflection, not just of light, but of heat when the sun is reflected from it.

Figure 10.3 *Grand Central Station, Birmingham.*

This proved to be the case with the showpiece building in Birmingham, the Grand Central Station (Figure 10.3), which has caused some discomfort on sunny days to nearby shop-keepers.

The contribution of Harry Brearley to the development and promotion of stainless steel has been referred to. In Sheffield, it is believed that it was he that created the industry based on this material and he is much celebrated. In 2013 a conference was held to mark the centenary of his first patent. The conference attendees all received a copy of a book that will have surprised many of them in the fair treatment it gives to the precedents and the other contributors to the industry.[6] Nevertheless, Sheffield claims stainless steel as its own. There is no clearer indication of this than the water feature that now graces the forecourt of Sheffield Midland Station. (Figure 10.5)

However, the Sheffield steel industry did build on the legacy of Brearley, especially in the development of the jet engine. Brearley's successor as head of research at the Brown-Firth laboratory, William Hatfield (1882–1943), collaborated with Frank Whittle (1907–1996) to develop many of the materials that were incorporated into the first jet

Figure 10.4 *Water feature at Sheffield Midland Station.*

engine. It was Whittle's hope that steels would have sufficient high temperature capability to get his engine off the ground and several were developed for engine rings (FV448, a Cr-Mo-V-Nb steel), engine discs (Jethete M152, a Cr-Mo-Ni steel), shafts (R20 19% Cr, 14% Ni, 1.7% Nb) and turbine blades (Rex 78, 14% Cr, 18% Ni, 4% Mo) all of which were used in the first engine to fly.[9] As we shall see later some of these materials were not strong enough at the operating temperatures and different materials had to be developed for eventual success.

Although nickel is not the dominant constituent of stainless steel, it is in the interest of nickel producers to promote the use of the material, given how large a proportion of the nickel produced it uses. So it was that in the 1970s the research laboratories in Wiggin Street set about seeking ways to promote the material.

One notable development was in finding a means of colouring the surface of stainless steel sheet. Two general approaches were researched, one chemical, the other physical. Both use the principle that the chromic oxide film on the metal surface is translucent. By varying its thickness, it is possible to produce, under controlled conditions, the colourful effect that is seen when there is an oil film on water.

The colouring effect arises because the incident light can then be reflected from the surface of the layer and from the interface between the layer and the substrate. If the two reflections are exactly out of phase, they cancel each other out. It is the thickness of the layer that determines which part of the spectrum is eliminated. The light coming from the surface will then have the colour that is complementary to the eliminated part of the spectrum. Computer users will be familiar with this effect. Where on the screen the light emitted is red, their printer will use an aquamarine ink to absorb the complementary colour to red which then shows red on the paper. Likewise, blue is produced in print by a yellow, and green by a magenta ink. Complementary colours are shown opposite each other in the colour circle in Figure 10.5.

The length of the light path through the surface layer varies with the angle of incidence so that the colour can be variable.

The thickness of the surface coating can be varied both chemically and when applying it by vapour deposition. By using the standard methods of screen printing, it is then possible to produce detailed patterns. Some examples are shown in Figure 10.6.

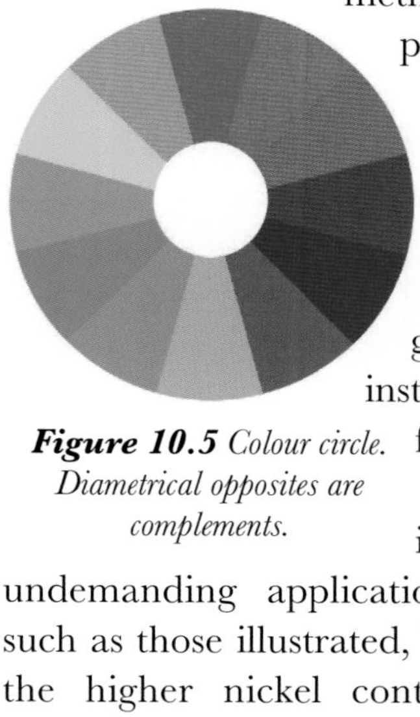

This method of decoration now finds extensive use, but since the appearance is obviously sensitive to the thickness and smoothness of the surface coating, it is not generally suitable for external use, unless, for instance, putting it behind glass. It is used widely for internal decoration and displays.

Figure 10.5 *Colour circle. Diametrical opposites are complements.*

The two thirds of the nickel produced that is used in stainless steel is mainly in undemanding applications, such as those illustrated, but the higher nickel content alloys, although overall being of more limited tonnages, perform in some of the most severe of environments. These are considered in the next chapter.

Figure 10.6 *Experimental samples of coloured stainless-steel. (Courtesy Peter Crouch)*

11

The Hardest Work

The most demanding of applications for structural materials are where the structure is subject to both high loads and corrosive environments. Corrosion is the result of both chemistry and temperature. The higher the temperature, the faster reactions occur.

In some applications, the temperature is cycled between high and low temperatures. This can exacerbate the problem of corrosion, because the material and the surface layer expand and contract, almost certainly at different rates, so that the corroded layer spalls off, or is cracked, revealing more of the metal surface to the environment.

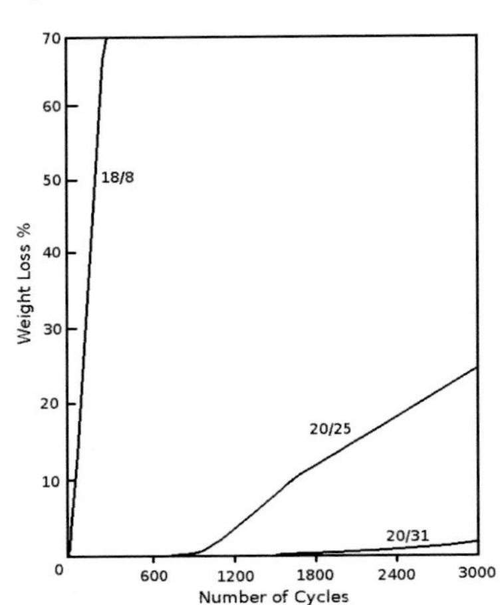

Figure 11.1 *Oxidation rate for three nickel containing alloys thermally cycled in air between ambient temperature and 980°C.*

The data shown in Figure 11.1 is from a paper by Herb Eiselstein, one of INCO's most noted alloy developers at its Huntington plant.[1] It shows, for several nickel-containing alloys, the weight loss due to oxidation when coupons of the material were cycled by being heated to 980°C for fifteen minutes and then allowed to cool in air for five minutes. In the diagram, the first number beside each curve is the nickel content, the second one, the chromium.

The common 18/8 steel performs very poorly in these circumstances, but an alloy like that used to contain the heating element in cookers, Inconel 800H, performs very well in this test. The slightly higher chromium is beneficial, but it is the good oxidation resistance of the substrate that has the largest effect.

More demanding are those applications where the material is under load and is also exposed to an aggressive environment. There are many different circumstances depending on the chemistry involved, the temperature at which the reactions take place and the pressures to which the material is subjected. In a majority of these conditions nickel alloys predominate in the structures involved.[2]

Much of the chemical industry takes it raw materials from oil refineries, which are themselves a major user of nickel alloys. Most of the range of nickel containing alloys will be found in a refinery, from the 18/8 variety to the high-performance alloys, which can approach 50% nickel content and even 100% nickel. There will also be other steels in the external parts of the complex structures that include pressure vessels, distillation towers and power generation plant. It would not be possible to operate the 600 or more oil refineries around the world without the nickel alloys.

Temperatures up to about 1000°C are found in an oil refining plant. The material used in its construction mostly starts out as tube or plate. For low temperature applications the material will be produced in tube or plate mills as wrought products, but for service above about 6-700°C the materials must be cast. For the straight tubes, this is done by pouring the molten alloy into rapidly spinning tubular moulds. This is the centrifugal casting process. For bends, of which there are many in a typical array (see Figure 11.2), the conventional casting processes are used, as they also are for other fittings, such as valve bodies.

Regardless of the temperature, the material is required to survive for 100,000 hours without breaking and in that time, it is limited to no more than 1% change in its dimensions. Slow deformation, called creep, occurs under load at high temperatures. In a large structure it can lead to failure, as the cumulative distortions of different elements transfer load to weak links in the structure.

Figure 11.2 *Centrifugally cast and conventionally cast high nickel alloys for oil refinery applications. (Products of Paralloy, a Doncasters Ltd company)*

Figure 11.3, which is adapted from data in a Nickel Institute publication, shows the failure stress for some of the high-performance alloys used in the chemical industry.

All the alloys represented in the figure have 25% chromium. The low carbon alloy had only 0.15% of carbon, the other two had 0.4-

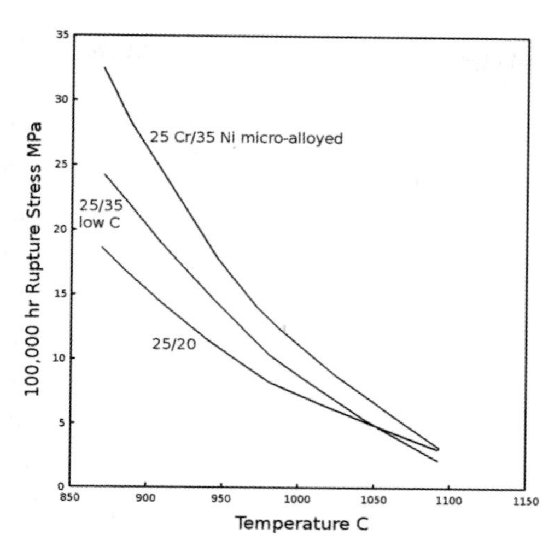

Figure 11.3 *The variation with temperature of the stress to cause failure after 100,000 hours for three high performance alloys.*

0.45%. It is evident that at moderate temperatures the higher nickel content gives the better resistance to deformation but when the carbon is used to form a fine dispersion of carbides of niobium, titanium and zirconium, in the micro-alloyed variant of the higher carbon alloys, it has a significant effect on strength over the whole temperature range. The higher nickel also confers the important resistance to chemical attack.

In addition to resistance to oxidation, materials in oil refineries are subject to carburising atmospheres and attack by hydrogen sulphide, which is formed from the sulphur that is commonly present in crude oil. Nickel has a very beneficial effect on resisting carburisation, but the higher nickel alloys do not perform so well when sulphur is a problem. One solution that is used involves lining the strong nickel alloys with weaker, but more sulphur-resistant alloys.

Figure 11.4 *A stress corrosion crack in an 18/8 stainless steel. (Courtesy John Oldfield)*[3]

The commonest form of failure under stress in corrosive environments is by stress corrosion cracking. If the chromium rich surface is penetrated, chemical attack can become localised and lead to the formation of cracks. A typical example is shown in Figure 11.4.

The crack both reduces the effective cross section of the material and intensifies the local stress, leading rapidly to failure, once the crack has formed.

Stress corrosion cracking is a common form of failure in high temperature applications and for many years the conventional wisdom was that stainless steels would not fail in this manner below 60°C.[4] However there have been several failures that have had fatal consequences at significantly lower temperatures. These all relate to one application where it should perhaps be recognised that nickel-containing steels should be avoided.

There have been several serious accidents at swimming pools, when the roof, or other items suspended on stainless steel wire or strip, have fallen. The crack shown in Figure 11.4 is from the investigation of one such accident. The authors concluded that it was the result of the environment in the swimming pool, where the chlorine used to disinfect the water combines with the bodily fluids that it is intended to disinfect, producing an atmosphere above the pool that contains chloramines (NH_2Cl and $NHCl_2$). These are obviously sources of both hydrogen and chlorine, and the known ability of nickel to remove hydrogen atoms from its compounds, almost certainly explains how highly acidic conditions are produced to cause and propagate cracks in the stainless-steel wires.

Following a number of these disasters the recommendation was to use a very rich stainless containing 25% nickel, 20% chromium and 7% molybdenum. This is detailed in a paper by Heselmans and Vermeij.[5] However a later report by Heselmans concerns a fatal event in which this alloy was the one that failed.[6]

This is a rare exception to the general rule that for demanding applications high nickel alloys are the materials of choice. The wide range of such applications is well illustrated in a Nickel Institute publication.[7]

There are other demanding applications to consider, but these use nickel alloys that are not among the family of stainless steels. They are described in Chapter 13.

12

Magnetism

Although magnetism was known to the ancients, and was used as a navigational aid, it was not until Faraday and Maxwell had clarified its intimate relationship to electricity that it found wider use.

In Faraday's first experiment in electricity he wound two coils of copper wire together around a stick, taking care to keep them insulated from each other. One coil he connected to a galvanometer and when he then linked the two ends of the other to the terminals of a battery the galvanometer needle flicked but did not continue to register a reading. He next wrapped one coil around a tube, inserted an iron pin into the tube and connected the coil to a battery. Leaving the coil connected, he withdrew the pin and found it to be magnetised. Inserting a second pin into the tube with the coil still connected, he then disconnected the coil from the battery and withdrew the pin, finding it to be magnetised in the opposite direction to the first pin. These simple experiments revealed much of what was needed to build electro-technology.[1]

The first consequence of the insight that had been gained was to enable electricity generation by moving a conductor in a magnetic field. The earliest generator used in industry was made in Birmingham and is now to be found in the Birmingham Science Museum (The Thinktank).

It was used by Elkingtons to provide the power for electroplating and was greatly approved of by Michael Faraday, when he visited

Birmingham for a British Association meeting. The magnetic material could only have been steel at that time.

The principle of transformation, changing the voltage of an alternating current, comes directly from Faraday's experiments, using input and output coils with different numbers of turns wrapped around opposite arms of a hollow core. This required the core to be made of an easily magnetised material, initially soft iron. Since Robert Hadfield's development of 3¼% silicon iron, which is as easy to magnetise but has a higher resistivity than iron, this has been the universal choice of material for transformation of mains electricity.

Iron has the virtue of being the cheapest magnetic material, but nickel and cobalt are also magnetic (as are some rare-earth elements, but only at low temperatures). However, iron can be magnetised more strongly than any other element, achieving magnetic strengths more than three times as high as that for nickel and 22% more than that of cobalt. All magnets lose their strength at high temperatures, but the range varies. For iron, it is up to 770°C, for nickel, 354°C, while cobalt retains its magnetism to the highest level, 1115°C.

As Guillaume demonstrated (Chapter 8), when the metals are alloyed there can be some interesting effects. The search for specific magnetic properties, which opened many applications for nickel, began in earnest after the invention of the telephone.

Alexander Graham Bell (1847–1922) is generally credited with this development, although there are other claimants. However, it was certainly at the research laboratories of the company that he founded that most of the alloys used in telephony were developed. The research was led by G W Elmen (1876–1957). His first alloy was composed of 80% nickel and 20% iron and was given the name Permalloy. This was used to wrap around the telegraph cable laid on the sea bed from New York to the Azores in 1924. It improved the signal-carrying capacity of the cable fivefold, making analogue (that is voice) transmissions possible. Subsequent improvements in the alloys used increased the efficiency dramatically, but later cables were of coaxial design rather than being of the shielded variety.

Nevertheless, the properties that Elmen sought have much wider application. Figure 12.1 shows the property that he measured across

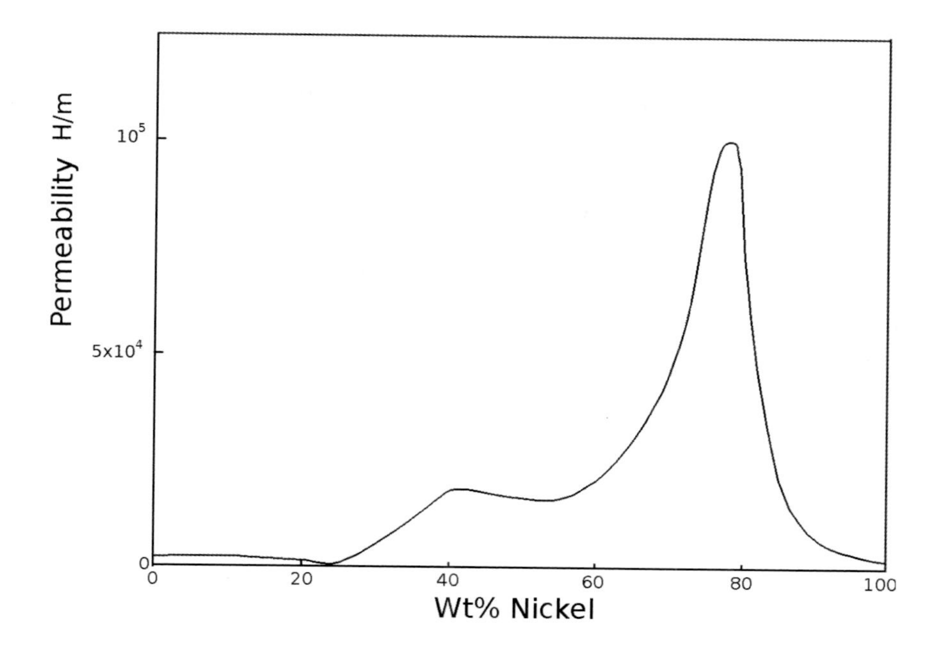

Figure 12.1 *The permeability of decarburised iron-nickel alloys.*

the range of alloys that Guillaume had studied. The ability of a material to concentrate magnetic flux is measured by its permeability.[2] Elmen originally identified the superior permeability of the alloy having approximately 20% iron and 80% nickel in 1914. At that time, pure elements, from which to make the alloys, were not readily available and all the materials were relatively impure. By the time that the cable to the Azores was constructed the properties had been improved to the level shown in Figure 12.1. This followed from the work of Yensen[3] at Westinghouse Electric. He had shown that taking pains to purify the alloy greatly improved the permeability, eventually identifying that annealing for a long time at a high temperature in hydrogen removed most of the carbon, which was responsible for much of the reduction in permeability. Using this treatment allowed commercial purity materials to be processed to yield the properties that Arnold and Elmen reported in 1923.

There were further developments at the Bell Laboratory by Elmen's colleagues, Cioffi, Kelsall, Bozorth and Dillinger,[4/5] when it was realised that reheating the decarburised alloy to above the Curie temperature and slowly cooling in a magnetic field led to a dramatic increase in magnetic properties. This was to have a major and lasting effect on all developments in magnetic materials, not only those used for their ease of magnetisation.

There are now many alloys based around the 80/20 Fe/Ni composition that are used as soft magnetic materials for specific applications. In some the composition was modified to improve the ease of producing the alloy in thin strip form – the most common format for the various applications – and some to increase the electrical resistivity, which is important for use in power transformation.

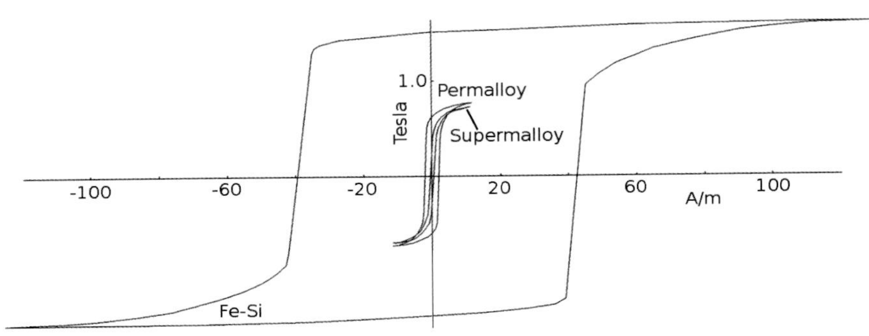

Figure 12.2 *Hysteresis loops for Permalloy, Supermalloy and silicon iron.*[6]

Were they less expensive, these materials would be the preferred choice for transformers, as may be appreciated by reference to Figure 12.2.

The hysteresis loops shown in the diagram are for Permalloy, an improved version of Permalloy, called Supermalloy, and the commonly used material for power transformers, 3¼% silicon iron. The curves show the flux density as it varies with the cycling of the magnetic field. The area contained within the curves represents the energy loss per cycle.

European electricity supply generally has a network of substations that transform the input 11,000-volt supply to the 400 volts of the local network and again to the 240 volts in the domestic unit. In each step the transformer has a silicon-iron core. It is the losses that these incur that make a major contribution to the fact that only about a third of the energy used at the power station to generate the electricity reaches the point of its application. Mains electricity operates at either 50 or 60 cycles per second. Although it is obvious that nickel-based alloys would do a better job, there is not enough nickel available for this application and it is too expensive.

The losses in the conventional transformer core material are more than a hundred times as high as those in the Supermalloy and these not only cause the evolution of heat but result in inertia. This is accepted in 50-60 cycle applications but for high frequency transformers and systems where stability of temperature is required, nickel alloy cores come into their own. There are many applications.

Among the many variants of Permalloy, Mu-metal is used for its high permeability at low field strengths. It is used to protect sensitive equipment from stray magnetic fields that could corrupt the signals. In Figure 12.2, the permeability is the initial slope of the B-H curve. When the material reaches its maximum flux density the permeability declines, so that shielding is only effective in low fields. To protect against strong fields, sensitive equipment must be enclosed in several layers of the soft magnetic material.

A once common application of these alloys in shielding was an envelope that surrounded the electron source in cathode ray tubes. This was to prevent stray fields from deflecting the beam and disturbing the signal, or picture, in the screen.

A piece of a ferromagnetic material that is demagnetised is composed of small regions, domains, which are small magnetic entities, each fully saturated, but arranged so that the magnetic fields are not in alignment. The domains can be observed in a polarising light microscope, because the magnetic field affects the plane of polarisation. An example is shown in Figure 12.3.

In the micrograph, the crystal grains can be seen to contain areas of differing contrast, corresponding to different directions of magnetisation.

Application of a magnetic field results in the favourably oriented domains growing at the expense of the others, until all domains are favourably oriented. The material is then magnetically saturated.

The high permeability materials do not achieve a higher level of saturation than the materials of which they are composed. It can be

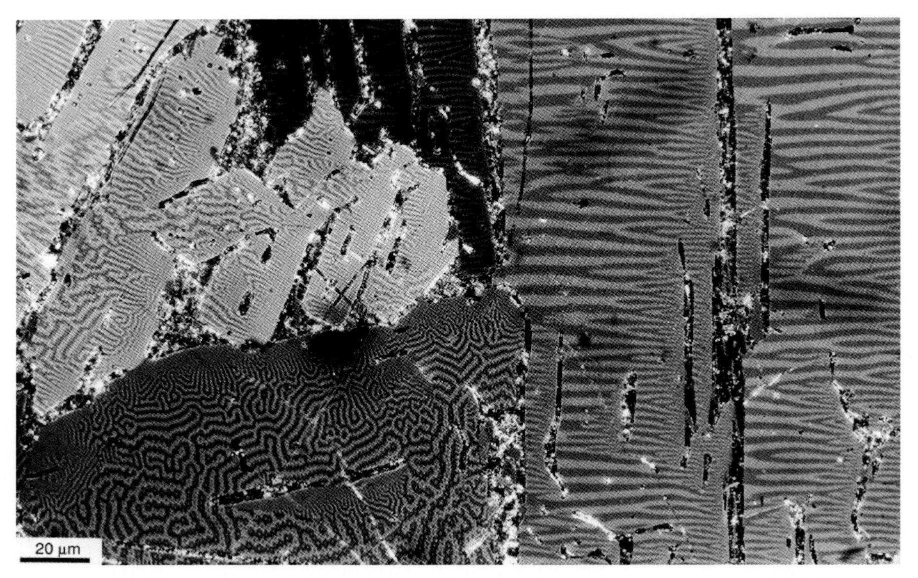

Figure 12.3 *Microstructure of a cast Fe-Nd-B magnetic material. (Courtesy Professor I R Harris)*

seen from Figure 12.2 that the Permalloy alloys saturate at around 0.6 or 0.7 Tesla. The saturation magnetisations at 20°C of iron, cobalt and nickel are, respectively, 2.16, 1.76 and 0.64 Tesla. The permeability reflects the ease of moving the domain boundaries, not the intensity of magnetisation in a domain. As shown by the observations of Yensen, that impurities reduce the permeability significantly, it is evident that the movement of domain boundaries can be impeded. The material is made harder to magnetise by the presence, especially, of second phase particles. (In Figure 12.3 the black and white features are inclusions, which are poorly distributed in the cast state but after processing are more uniformly and finely dispersed to make the iron-neodymium-boron material a hard magnet).

Materials like Permalloy, Supermalloy and Mu-metal are notable examples of soft magnetic materials. They would be no good for making compass needles or fridge magnets. For these it is desirable to have a magnetic property that resists decay. Hard magnets are a mostly unseen but essential part of everyday life. Audio equipment requires speakers, which all contain hard magnets, car doors are locked and unlocked by magnetic forces and car windows wind up and down using motors that are operated by hard magnets.

There have been hard magnets for many years. The carbide particles in steels resist demagnetisation and bar and steel horseshoe magnets, made for many years for simple tasks, were carbon steels. It was not, however, until something more substantial could be produced that the magnet became an essential ingredient of modern technology. The first strong magnet was, of course, a material that contained nickel.

T Mishima (1893–1975), working at Tokyo Imperial University in 1931, discovered that the nickel-iron composition, which Guillaume had shown has its Curie point at or around room temperature, has the magnetism restored when aluminium is added.[7] It forms a hard-magnetic compound that found early use in loud speakers. Its utility was greatly enhanced by replacing some of the iron with cobalt and by magnetic annealing. Thereafter, magnets were available that could fulfil serious engineering functions. (Hard magnets are discussed further in Appendix 1.4).

They could also contribute to entertainment, as an article on guitars illustrates.[8] Prange notes that the electric guitar was included in Benny Goodman's band in 1939 (played by Charlie Christian), but it was not until 1947 that the magnet used to contribute to the amplification was a permanent magnet made of the AlNiCo alloy, which was the ultimate outcome of Mishima's work. The delay in using this much lighter and more portable device was because of nickel's strategic importance in the 1939–1945 war.

Apart from the renewed importance of nickel in armour plate production, and its use in aircraft engines, the most important application of nickel during the war was a new contribution from Birmingham to the story of nickel.

Randall (1905–1984) and Boot (1917–1983) did not invent the magnetron, but they were responsible for making it possible to develop radar by increasing the magnetron's power output and generating short wavelength microwaves. Electromagnetic waves, including light, can detect nothing that is smaller than their wavelength. So, to detect small details, such as the periscope of submarines, that would give away the presence of an enemy, it was necessary to have radiation of a wavelength shorter than the dimensions of the object. Something less than 10 cm was required. Also, the power output had to be high to give a strong reflected signal.

These were the problems that Randall and Boot solved in the Physics Department of Birmingham University.

The key requirements were in the geometry that produced resonance at the short wavelength and a magnetic field that gave a strong enough output signal. Randall and Boot developed the cavity magnetron, which controlled the wavelength range. The geometry was in the anode block; (an original is now in the Science Museum in London, several others are held at Birmingham University). They did not need to solve the problem of creating a strong magnetic field; in the laboratory they could use electromagnets. However, to install it in practical portable devices it was necessary to have strong permanent magnets and it was the availability of magnetically annealed AlNiCo that made airborne radar possible.

Figure 12.4 *An early magnetron made in the workshop of the Physics Department at Birmingham University, showing the anode form.* *© The University of Birmingham Research and Cultural Collections.*

In the cavity magnetron, there is a heated cathode projecting into the central hole in the anode. The cathode is like that in a triode; it has a coating of a material that easily emits electrons, heated by a nickel-chrome element. Pulses of power applied to this cause streams of electrons to connect with the thinnest parts of the anode, causing vibration that results in the emission of microwaves. The electron currents go variously to different sections of the anode. The microwaves are directed by a strong magnetic field from horseshoe AlNiCo magnets that surround the anode and with a diameter of the anode. The microwaves are collimated by a wave guide and are directed parallel to the magnetic field as a beam.

Radar devices now use other methods more commonly than magnetrons, but more magnetrons are made than before, because they are the source of power for the microwave ovens that are now found in most kitchens (Figure 12.5).

Figure 12.5 *A sectioned magnetron from a microwave oven.*
© The University of Birmingham Research and Cultural Collections.

In Figure 12.4 the magnets would have been of the horseshoe type and the magnetic field caused the microwaves to emerge along the long axis of the figure. With more powerful magnets that could be moulded as rings, a more compact configuration is possible, as shown in Figure 12.5. The microwaves are emitted along the vertical axis in this example, which is a sectioned magnetron from an oven made around 1990.

In an oven the microwaves are not collimated to form a beam, as would have been the case for radar, but scattered to fill, as far as possible, the cooking space. There the fluctuating magnetic field causes eddy currents in conductive materials, which in turn causes internal heating of the material.

The magnetic properties obtained by Mishima were significantly better than those of the existing steel magnets. Magnetic annealing raised the saturation energy to about three times that of steel, making the development of radar possible. Many other applications followed, recording heads, electrically amplified musical instruments, brushless motors and many more, but the search for higher power, which permits lower weight, continued. By control of the manufacturing process it was possible to achieve higher magnetic strengths. This is because the magnetic flux density is different along different crystal directions. Techniques to align the easy direction of magnetisation in the crystals with the magnetising field raised the strength further. This is shown in the bar chart in Figure 12.6.

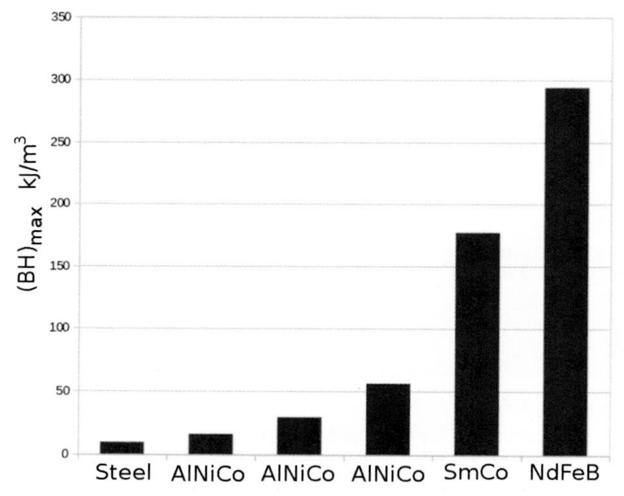

Figure 12.6 *The magnetic energy, (BH)max for several materials. The AlNiCo material was progressively enhanced by developments in processing but was overtaken by stronger magnetic materials.*

Both soft and hard magnetic materials have been improved by control of the process of manufacture. In the case of the soft magnetic materials, the discovery by Yensen,[3] that the removal of carbon improved the permeability of the iron-nickel alloys, illustrated that second phase particles have the effect of impeding domain boundary movement. This is also a clue to how a magnet may be made hard. For a soft magnet, a single-phase material is required, but the best properties that can be obtained are not in pure metals. Supermalloy is composed of 75% nickel, 20% iron and 5% molybdenum. The addition of molybdenum reduces the saturation magnetisation relative to the 75/25 composition (Permalloy) but improves the initial permeability. The effect of magnetic annealing on permeability is thought to be by causing atom pairs to become aligned in a favourable configuration through the effect of the applied magnetic field acting at temperatures at which the atoms are still mobile.

The strongest magnets now available are based on iron, with alloying additions of neodymium and boron. These are found in most electrical motors, especially those used in portable devices and in power applications. Wind turbines contain large quantities of these magnets in the generators that convert the rotary motion of the shaft connected to the blades to the output electricity. There is no nickel in the magnets, but all of them are electroplated with nickel because the corrosion resistance of the alloys is extremely poor.

13

Power Generation

a) Jet aircraft

Nickel is the most important material in the development of jet flight, in the creation of a gas turbine with enough power to fly an aircraft.

The first gas turbine to generate useful power was developed in Norway by Æ. Elling (1861–1949) in 1903. He recognised that the way to increase the power generated is to increase the operating temperature of the engine. He also proposed the configuration that is used today in jet engines; however the materials to make an engine powerful enough to get an aeroplane off the ground did not exist during his career.

The basic principle of a gas turbine is contained in the gas law, the combination of Boyle's Law and Charles' Law:

$$PV/RT=\text{Const.}$$

Where P is the gas pressure, V is its volume, R is a universal constant and T is the absolute temperature. Taking room temperature to be 293K, this means that if unit volume is heated to 586K it will exert twice the pressure. Similarly, if unit volume is compressed to half its volume it will become twice as hot (neglecting heat losses).

The gas turbine operates to raise the pressure in two stages. Firstly, a compressor increases the pressure of the air passing through it; the oxygen in the air is then burned with a fuel to further increase the temperature and therefore the pressure, which is then released through a turbine section, connected on the same shaft as the compressor so that the turbine provides the energy to work the compressor. In a pure jet engine, the remaining energy in the emerging hot gas pushes back on the structure of the aircraft to propel it forward.

Unlike a propeller driven aircraft, which operates by pushing against the surrounding air, the density of the air is a factor in the performance of the engine only insofar as air is needed for the machine to work. For this reason, jet-powered aircraft can fly at higher altitudes than propeller aircraft.

Frank Whittle (1907–1996) is deservedly credited with inventing the jet engine, but the lack of support for his proposal within the Ministry of Defence led to his patent being published, making full knowledge of his design available to Hans von Ohain (1911–1998). The two men were then in a race to get a jet powered aircraft flying. Ohain won that race; Whittle achieved flight two years later but neither of their aircraft contributed to the war effort. In both cases this was because of a lack of the alloys necessary for the most critical parts of the engine.

In Ohain's case it was because Germany's limited supply of nickel was needed for making armour plate, in Whittle's case it was because he thought that steel would be good enough and the work to supply a material with higher temperature strength did not start until he, like Ohain, found that the available steels were inadequate. The Tinidur used by Ohain had a better chance than the Stayblade originally used by Whittle. Tinidur was a steel with 36% nickel and 15% chromium, Stayblade, only 8.5% nickel.[1] Both had an addition of titanium to make use of the Ni_3Ti precipitation hardening following the work of Pilling and Merica.[2] Whittle's first flight engine used a stronger steel, but none of the steel-based alloys had sufficient creep strength and the distortion of the turbine blades limited very severely the operating time of the engine. Something capable of withstanding higher temperatures was required.

Both Whittle and Ohain used a centrifugal compressor, based on a turbocharger design. In both designs a double entry – that is two-sided – impeller was used to increase the incoming air pressure, forcing it into combustion chambers arranged around the circumference of the machine. Whittle's machine is illustrated in the semi-schematic diagram shown in Figure 13.1.

The diagram shows the impeller connected to a disc, on which is a turbine blade – marked as a rotor blade. There is another airfoil labelled NGV, which stands for nozzle guide vane. The NGV is fixed to the engine casing and has a form that deflects the hot, high pressure gas leaving the combustion chamber onto the turbine blade, forcing the disc to rotate and drive the impeller, which in turn is compressing the air into the combustion chambers. In this design the energy extracted from the gas stream is only enough to generate the required compression of the incoming air. The remaining energy is concentrated in the emitted gas stream, the jet, from which the machine gets its popular name.

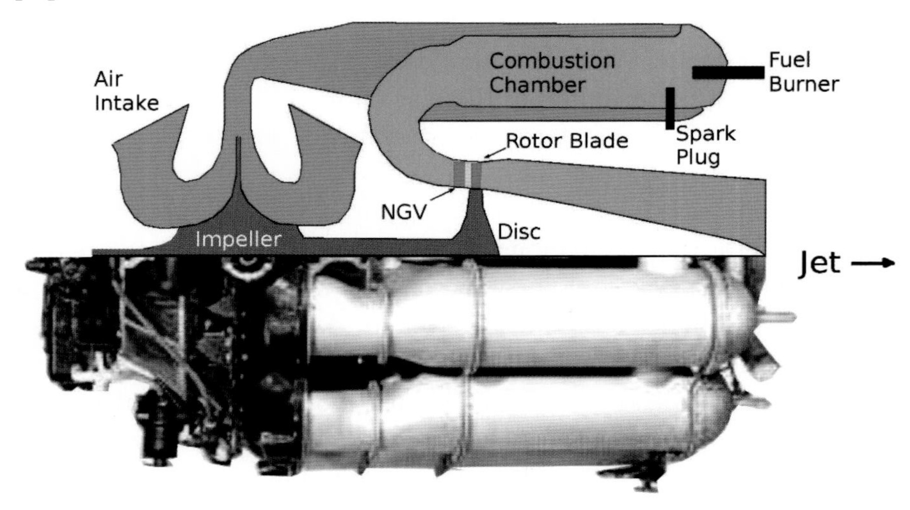

Figure 13.1 *Semi-schematic diagram of the original Whittle engine.*[3]

The critical components are the containment for the combustion chambers, the NGVs and the turbine blades. These must withstand the highest temperatures. Especially the airfoils must be highly creep resistant. Any change of shape from that intended will reduce the efficiency of the engine and, in extremis, could cause it to stall. The German engine is said to have had the airfoils replaced frequently.

In Whittle's case, having belatedly obtained Government support from Henry Tizard, a project was established to develop a better material. This task was given to the Mond Nickel Research Laboratory. Experimental materials were creep tested at the Royal Aircraft Establishment, but the alloy development was in Birmingham.

In 1939, when the need was recognised, L.B. Pfeil (1898–1969) was assistant director of the Mond Laboratory. He and the director (W.T. Griffiths) set the aim of developing an alloy that could resist a stress of 3 tons per square inch at 750°C for 300 hours and suffer a permanent strain of no more than 0.1%.[4]

The first alloy developed was given the designation Nimonic 75. It subsequently found use as a sheet alloy, used in combustors in the early engines. It contained 5% of iron – it is always attractive to try to include the least expensive alloying elements – but it had insufficient creep resistance for turbine blades. The first successful Whittle engines had an iron free alloy, more closely related to the 80/20 nickel-chromium composition with additions of titanium and aluminium to provide extra strength.

It did not require much insight into the properties of the various alloy systems to appreciate that the place to start the development was to use alloys based on the material that was well established for resistance heating: 80% nickel and 20% chromium. It was obviously corrosion resistant and capable of sustained high temperature service.

The compositions of these and subsequent Nimonic alloys are illustrated in Figure 13.2.

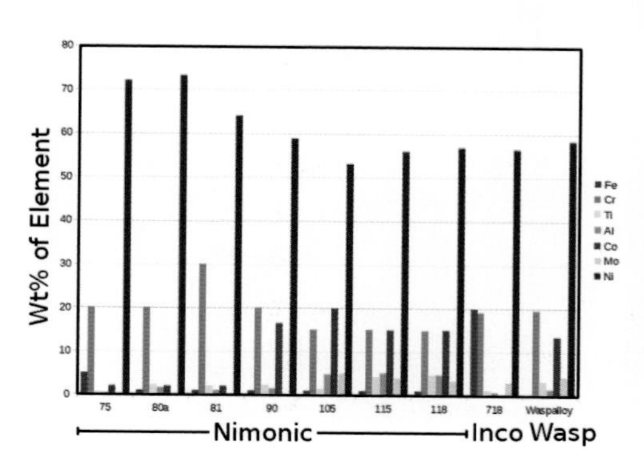

Figure 13.2 *Compositions of the Nimonic alloys used for turbine blades in early jet engines. The two most important disc alloys, Inco718 and Waspalloy, are also shown.*

Credit is usually given to Leonard Pfeil (Figure 13.3) for the development of the Nimonic alloys, but it must be recognised that he was the senior manager in the laboratory. His name is, indeed, on the patent,[5] but the foot soldiers of the laboratory undoubtedly made significant contributions. It is probable that one of the co-editors of the book 'The Nimonic Alloys', Walter Betteridge, had much to do with the development.[6]

Figure 13.3 *Leonard Pfeil (1898–1969)*
Assistant Director of the Mond Nickel Laboratory.

Later Nimonic alloys contained cobalt and molybdenum which raised the temperature capability as can be seen from Figure 13.4.

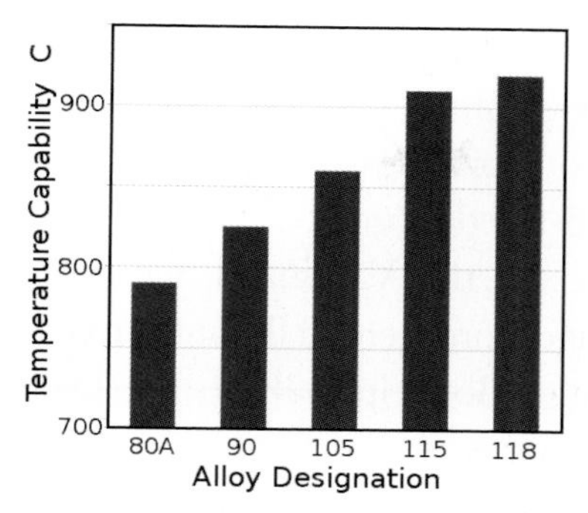

Figure 13.4 *Temperature capability of the wrought Nimonic series of alloys for 1000hr life at 138 N/sq.mm.*[7]

These alloys served the development of turbine powered flight for several decades. Their use expanded significantly when the Whittle engine design was replaced by the configuration that was first envisaged by Elling, for which the detailed design was developed by A.A. Griffith (1893–1963) in 1926. Griffith had proposed that the turbine use an axial compressor and should replace the existing piston engines to power a propeller. To do this the configuration is designed to extract the maximum energy from the hot gas, unlike a pure jet, where most of the energy goes into the jet stream that provides the motive power, see Figure 13.5.

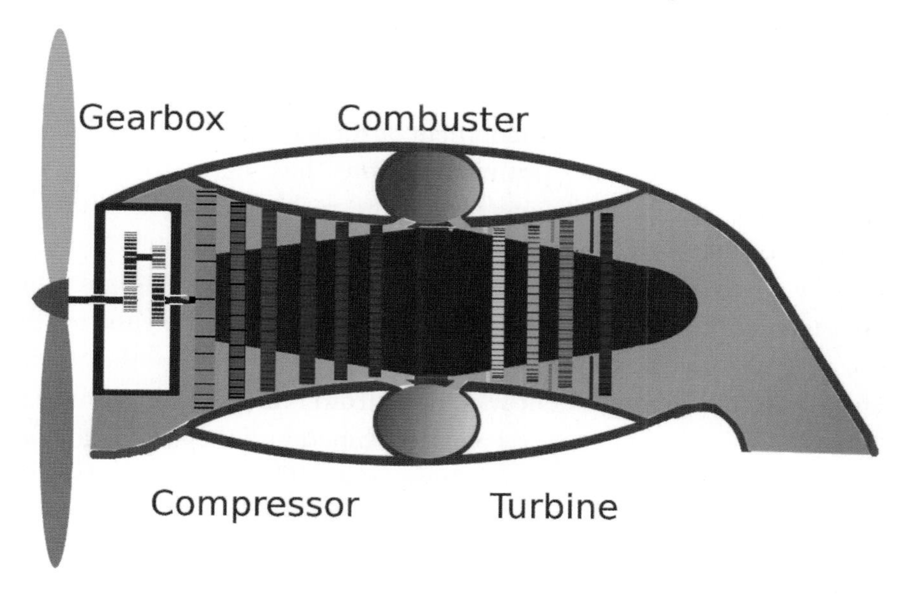

Figure 13.5 *Schematic diagram of the configuration for a turboprop engine. Several stages of turbine blades and nozzle guide vanes are required to extract the energy from the gas stream.*

Whittle's impeller design offered little scope for development, while the axial compressor has evolved extensively since it was first used for commercial powered flight in 1948 in the Rolls Royce Dart engine. The aircraft was the Vickers Viscount. A similar axial compressor was also used in the first pure jet engine to power a commercial aircraft which, like the Whittle engine, had just one row of nozzle guide vanes and one of turbine blades. The aircraft was the ill-fated Comet.

The Viscount and its larger stable-mate, the Vanguard, were the work-horses for much, relatively short range, commercial flight for many years and they used a lot of the Nimonic alloys, initially in the Dart engine, but later versions used the Tyne engine.

Air contains a little over 20% of oxygen, and it is the reaction of this with the kerosene fuel that generates the high pressure in the gas leaving the combustor. If all the oxygen is used up in burning the fuel, a temperature of close to 2000°C can be achieved. The hot gas emerging from the combustion chamber impinges first on a row of nozzle guide vanes, which must be able to resist deformation at the operating temperature. As the gas passes through the turbine section its temperature falls, and the successive rows of turbine blades are made progressively larger to extract the greatest possible quantity of energy from the stream.

There has been much development of the jet engine but to lift larger and larger aircraft the principal means has been to raise the temperature at which the gas enters the turbine section. The first engines to exceed the temperature capability of the wrought Nimonic alloys were the Conway and the Tyne. The turbine inlet temperature for these engines was over 1000°C. The hottest components are the nozzle guide vanes at the entry to the turbine. To achieve the service life required, this component was made from Nimonic 108 (similar to 105) and was cooled by air passages machined into the airfoil. The air was supplied by a bleed from the compressor. This was the first use of cooling.

It was also the last use of wrought alloys for the hottest components. In degrees-absolute, 1000°C is 1273K. Nickel melts at 1728K, but, as can be seen from Figure 13.6, the turbine inlet temperature was pushed ever higher in subsequent designs. The initial flat portion in the curve for the cooled blades was the last use of wrought alloys for the nozzle guide vanes. Thereafter cast materials were used, and subsequent developments in the metallurgy of the cast alloys, increasingly complex cooling geometries and the use of thermal barrier coatings have led to turbine inlet temperatures significantly higher than the melting temperature of the alloys used. Nevertheless, whether cast or wrought materials were used, the turbine section was, and remains, almost wholly nickel based.

The first wave of engine design was concentrated on engines that derived their thrust from the hot gas stream. The second adopted a different concept, first proposed, again, by A.A. Griffiths. The principle is illustrated in Figure 13.7.

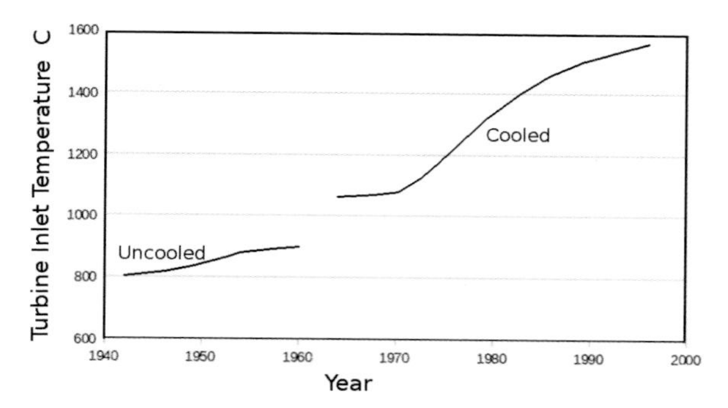

Figure 13.6 *The progression of the turbine inlet temperature through the second half of the 20th century.*

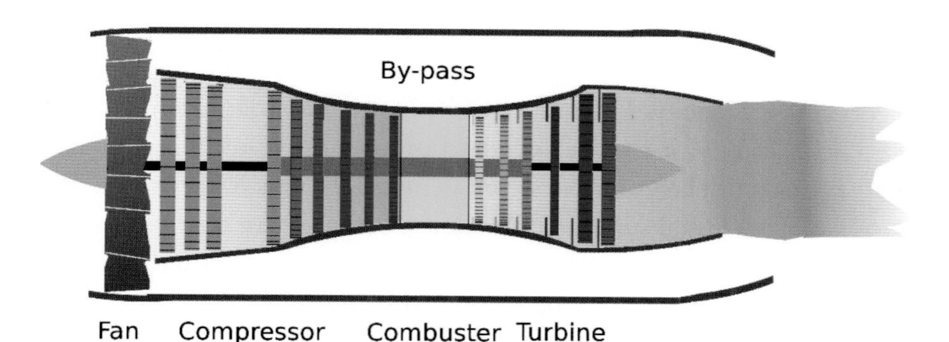

Figure 13.7 *Schematic diagram of a turbofan engine.*

The engine has two concentric shafts (except Rolls Royce engines, which since the first variant of the RB211 engine commonly have three concentric shafts). The one linking the first set of compressor-airfoils to the last group of turbine-airfoils is referred to as the low-pressure spool; the shaft passes down the centre of a tubular shaft that links the high-pressure compressor section to the high-pressure turbine section. In this design the high-pressure section forces the air into the combustor, while as much energy as possible is retained to drive the low-pressure shaft, which has a large set of blades, called a fan, that pushes air down a by-pass duct. This cool air surrounds the emerging hot gas. Most of the aircraft thrust comes from the fan. A turbo fan engine can achieve higher speeds than one that drives a propeller but not as fast as a pure jet engine. It is the most efficient available design and the quietest. It also uses more high temperature alloys.

b) Power Generation

The Industrial Revolution was powered by coal. It was burnt to raise steam to work machinery and to generate electricity. The energy that is generated all derives from the reaction of carbon with oxygen from the air, obviously resulting in a large amount of carbon dioxide being emitted, as well as the large waste of energy that is shown, in part, by the steam rising from cooling towers. When gas is burned, the methane, CH_4 results in twice as much water vapour as carbon dioxide, which is obviously a better outcome. The energy per unit weight of the fuel is also higher. Pure carbon generates 34080 kilojoules per kilogram, slightly more than the best coal, while methane produces 55,530 kilojoules per kilogram.[8] There are many more advantages to using gas to generate electricity. The unit can be placed close to the demand and sized accordingly; it can be switched on and off if required, with a relatively short delay – unlike steam generation, which needs to boil a lot of water and then superheat the steam before it can generate power – it is clean and more efficient. The efficiency can be further raised by using a combined cycle configuration in which the waste heat in the gas stream is used to power a secondary steam turbine.

Early gas turbines were single spool machines in which all stages were designed for optimum performance at a fixed output, but the tunable aircraft technology has now penetrated the industry, so that whereas fixed geometry turbines would use only stainless steel airfoils, it is now the case that the turbine section is based on the technology developed by the aircraft industry. The alloys used in the airfoils are not only nickel based, they use the advanced metallurgy that has been developed to get the most out of the materials.

Some of the methods used are illustrated in Figure 13.8.

In advanced designs, the cooling geometries have become very intricate. They are produced by pouring liquid alloy into a mould that defines the external form of the product and contains a ceramic core; an example is shown in the figure. After the metal has solidified, the core is removed using strong alkaline solutions that do not attack the metal but dissolve the core.

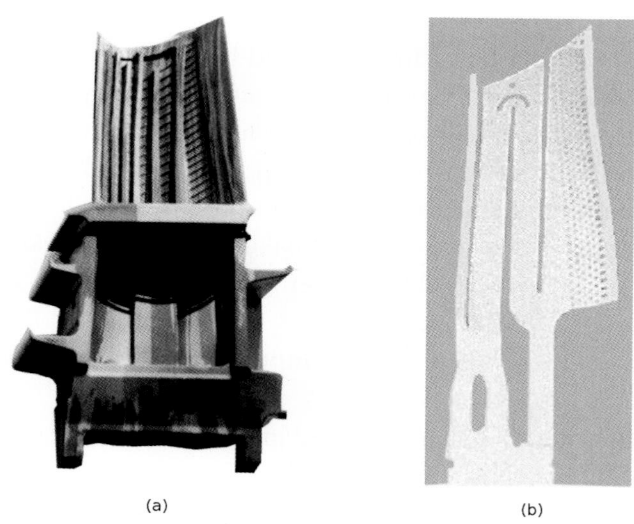

(a) (b)

Figure 13.8 *An industrial gas turbine airfoil, sectioned to remove one surface, to reveal the arrangement of the cooling channels within. The channels result from the casting being made around a ceramic core like the one on the right.*

A further development is to ensure that the solidification proceeds from the bottom upwards. As is revealed by the etchant used on the airfoil, this causes the grain structure to be not only coarse but elongated in the direction of growth. In the most advanced cases it is arranged that only one grain fills the whole of the casting.

It was shown in Figure 7.7 that metals are stronger and tougher when the grain size is small. This is true only at low temperatures. A rule of thumb is that coarse grained structures are stronger than those with fine grains above half the melting temperature. The hottest components are working at temperatures very close to the melting temperature and here the ultimate large grain size, just one grain, has the greatest strength.

The Nimonic alloys continue to be used in the lower temperature areas of the turbine. In the Olympus engine, used to power the Concorde, they were even used in the highest-pressure compressor blades. In the Olympus, the air entered the combustor at 600°C, too hot for steels to handle and the last few rows in the compressor were made of Nimonic 90. The turbine nozzle guide vane was, like Nimonic 90, an alloy developed at the Mond Nickel Laboratories. Stuart Shaw was its developer, as he was also for Inco 939, a workhorse alloy used in industrial gas turbines. 939 is a high chromium and high cobalt alloy specifically designed for the long service life required in the industrial application.

Although only a small part of total nickel usage, the gas turbine industry is perhaps the most important application in which nickel is the most essential ingredient. No other material has been found that can deliver the benefits that nickel offers, making two of the most important requirements of modern life, power on tap and cheap air travel, dependent on its availability.

The structure of these high temperature alloys is discussed in Appendix 2.7.

The highest temperature achieved in cast airfoils is by using several hardening phases, $Ni_3(Al,Ti)$, (titanium and aluminium are interchangeable in the γ'), and Ni_3Nb, which also has a degree of coherence with the nickel-rich matrix, although its structure is slightly tetragonal. (Niobium is a constituent of Inco 718, the most used of all high temperature aerospace alloys, which was developed by Clarence Bieber at INCO's US laboratory).[9] Matrix hardening elements are also used; cobalt and molybdenum are the most significant. There is always some chromium, but the dominant element continues to be nickel.

14

The Strongest Steels

It is surprising that the strongest steels, which make use of hardening by intermetallics similar to γ', were not developed until a decade after the Nimonic alloys. The age hardening effect associated with alloys that contained both nickel and aluminium (or titanium) had been known since the work of Merica on the copper-nickel alloy, Monel.[1] It was not until 1963 that the structure of the hardening phase was identified.[2] This was at INCO's US laboratories, after the development of electron microscopy. However, the strongest steels, the maraging steels were developed before this detailed insight was available. The developer responsible was Clarence Bieber, one of the most prolific alloy developers, working at the INCO research laboratory in New Jersey.[3]

The discussion of strengthening mechanisms in Appendix 2 concentrates on the interaction of nickel with carbon in the development of the strength and hardness in steels, so it may seem strange that the strongest steels are essentially carbon free. In carbon steels, it is the distortion caused by the carbon that is trapped in the iron lattice on rapid cooling, and its precipitation in the iron lattice on tempering, that gives rise to the strengthening effect. The various heat treatments used to develop the properties of iron-carbon alloys were simple enough to have been discovered by chance. This is not the case for maraging steels. Indeed, had the X-ray analysis of alloy structures not been developed, it is unlikely that the more complex treatments necessary to develop maraging steel properties would have become known.

The maraging steels do not only have the highest strengths that are available for significant structures, but they also have good toughness and fatigue resistance, and, having a high nickel content, typically 18%, they have better corrosion resistance than other high strength steels.

Figure 14.1 *The Ordsall Chord Bridge over the River Irwell in Manchester erected to link the two main railway stations in the city.*

In chapter 7 the amount of a simple steel required to build the first important wrought steel structure, the Forth rail bridge, resulted in a heavy structure with a lot of cross-bracing, as seen in Figure 7.3. The development of higher strength steels eventually enabled a lighter structure, of essentially similar design, to be erected over a river in Quebec, and today it is possible to create truly elegant structures with strength provided by maraging steels, as shown in Figure 14.1.

In the picture, the lattice work of tie bars is composed of pre-stressed maraging steel. High strength permits structures to be made much lighter. This is useful in applications such as aircraft and rocketry, where lightening the structure permits an increased payload. Maraging steels have also been used for ultra-centrifuge bodies to reduce the inertial load.

One of the most significant uses of maraging steels is for some of the most highly stressed parts in the undercarriage of large aircraft. In this application, the strength permits lower weight, but these structures also must withstand the repeated impact loading on landing, requiring good toughness and fatigue strength.

One application for maraging steel that did not get made was the big gun that Saddam Hussein intended to have built to threaten his distant enemies.[4] For better or worse, modern warfare is now conducted by other means, so that cannon development, which motivated Henry Bessemer to develop his steel making process, is a thing of the past.

The maraging steels have a typical composition of 18% nickel, 8% cobalt, 5% molybdenum and up to 2% of titanium. Bieber's original work used 25% of nickel, but for reasons that will become clear the lower level is preferable. Carbon is kept as low as possible.

As can be seen from Figure A.5, nickel extends the temperature range of the face centred cubic phase, austenite. This is different from cobalt and molybdenum, which both restrict the austenite range. The binary phase diagrams are shown in Figure 14.2.

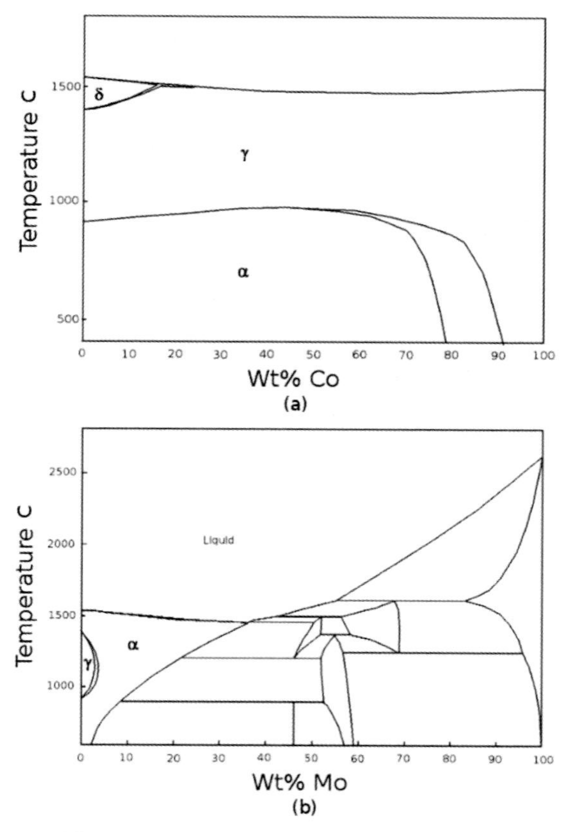

Figure 14.2 *Phase diagrams for a) iron-cobalt. b) iron-molybdenum.*

Over the composition range of interest, cobalt has little effect on the temperature at which austenite transforms to ferrite under equilibrium conditions, while molybdenum, a strong ferrite-forming element, restricts the austenite range very severely. However, achieving equilibrium requires much slower cooling from the high temperature austenite than in an iron carbon alloy. The small carbon atom diffuses in the iron lattice much more quickly than the much larger nickel, cobalt and molybdenum atoms. With the austenite stabilising effect of the nickel and the ferrite stabilising effect of the molybdenum cancelling out, the high temperature austenite phase can easily be retained at moderate cooling rates following annealing. The annealing temperature is typically 820°C.

In this condition the material is relatively soft. It becomes plastic at a stress of about 800MN/m^2 and can be machined to final form in this condition. To develop a much higher strength, the component is annealed at 480°C. Depending on how long it is annealed and how much titanium is present in the alloy, strengths between 1400 and 2400MN/m^2 can be achieved. There is also some effect of section size, which affects the rate of cooling from the annealing temperature. For comparison with these properties, a soft mild steel would become plastic at about 300MN/m^2, while the highest strength conventional carbon steel, containing chromium, molybdenum and nickel (AISI4340) can reach loads of 1800MN/m^2 before deforming permanently.

An interesting and useful feature of the change that occurs on ageing a maraging steel is that it causes no distortion. This is not the case for carbon steels. The possibility of making the component to final form when the material is in its softest condition is a significant advantage.

Carbon-free martensite is discussed further in Appendix 2.8.

15

Portable Power and Energy Storage

The first source of generated electricity was portable, at least in principle. Volta's cell could have been carried from place to place, but the ability to generate electricity from magnets and movement soon replaced the crude early cells. It was the widespread use of radio that created a demand for a portable source of power and that was in the form of rechargeable lead-acid batteries. Until about the same era, cars had been started using a spark generated by cranking the engine. The lead-acid battery became truly portable as it took its place under the bonnet of cars to enable a dynamo to save the effort of cranking. Portable it may have been, but the lead-acid battery was far too heavy to create a market for convenient devices to be carried around.

The real advent of portable power required the invention of the semi-conductor. Circuitry based on the semi-conductor required lower power, took up less room and so resulted in portable radios and the Sony Walkman. The power source for any such device was initially provided by small batteries in which the reaction is between zinc and manganese oxide with an ammonium chloride paste as the electrolyte, generating 1.5 volts. These are referred to as zinc-carbon batteries. The battery cannot be recharged and the risk of running out of power at inconvenient times soon led to the power being supplied by rechargeable batteries.

There have been many chemical combinations devised to provide power from electro-chemical sources; the lead-acid battery was for many years the most important. A single cell of such a battery generates 2.3 volts but the energy density is quite low. Nickel based batteries generate about 1.2 volts. The first rechargeable battery to dominate the market had a cadmium anode and a nickel oxide hydroxide (NiOOH) cathode; the electrolyte is potassium hydroxide. The power density is lower than that of the zinc-carbon battery, but the small closed cells are of the same size and shape as the non-rechargeable type. These are manufactured like Swiss rolls by layering cadmium foil, a paste containing the electrolyte, a lightweight porous-nickel foil filled with NiOOH and an inert separating layer, and then rolling them into a form that fits into the outer case. The nickel within the NiOOH is connected to the cell base, the cadmium to the positive terminal. The arrangement is shown in Figure 15.1.

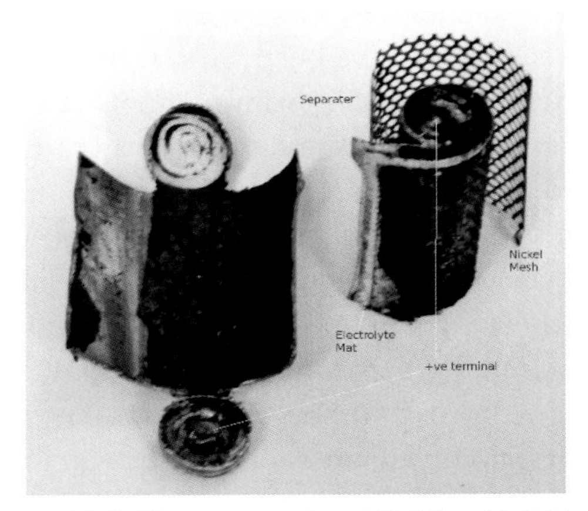

Figure 15.1 *The structure of an NiMH or Ni-Cd battery.*

In the discharged condition the four layers are as follows: the $Ni(OH)_2$, which is held in place by, in this example, electroformed nickel mesh – the nickel act as the current collector; the next layer is a fabric wetted with potassium hydroxide, separating the $Ni(OH)_2$ from what is now cadmium hydroxide. The last of the four-layer sequence is an impermeable barrier. While the nickel cadmium battery is supplying power, its cadmium negative electrode is being converted to cadmium hydroxide, while the NiOOH positive electrode in combination with the water is generating the hydroxyl ions and reducing to nickel hydroxide. In the small batteries, the nickel oxyhydroxide, the active mass, is held in place, as shown, by a nickel foil that supports the whole structure.

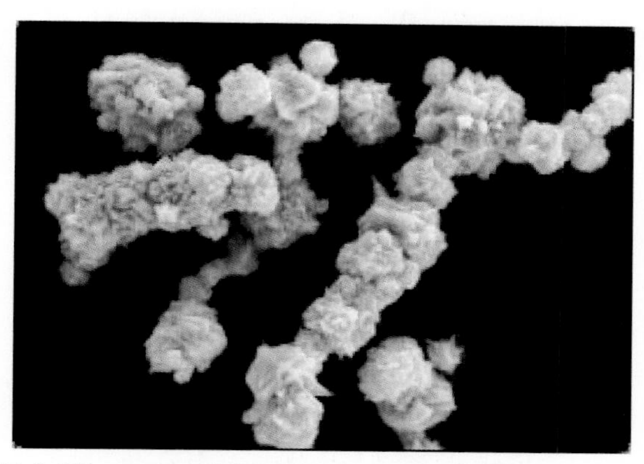

Figure 15.2 *Filamentary nickel powder, Vale-Inco 255. (Courtesy Tony Hart)*

For larger batteries, the NiOOH active mass is loaded into a different form of lightweight porous nickel. Such structures are referred to as foams. The initial method of making the foam was by sintering a nickel powder like that shown in Figure 15.2.

The powder is produced by the Mond carbonyl process. As related in Chapter 2, the refining of the nickel in the carbonyl process is by vapour transport, culminating in vapour deposition that makes quite regular spheres. To use the process to make the filamentary powder shown in the figure, the depositing conditions are modified by injecting low levels of other selected gases into the chamber. The development of the processes was at INCO's refinery in Clydach. The processing by sintering, to determine how low a density could be achieved while creating a useable product, was carried out at the laboratory in Birmingham.

The low-density sintered product was overtaken by a nickel foam of even lower density, developed by the Japanese battery manufacturer, Matsushita.[1] The process used was to apply a coating of nickel, by electroplating, onto a polyurethane foam. This produced densities as low as 3% of solid nickel, or to put it another way, 97% of the volume was available for containing the active mass.

The response of INCO was to build a plant in which the nickel was introduced as unrefined matt, and polyurethane foam is coated with refined nickel using the carbonyl process.[2] This less expensive method can achieve results like the electroplated product. It involves fewer processing stages and generally cleaner procedures. The nickel foam produced by INCO is illustrated in Figure 15.3.

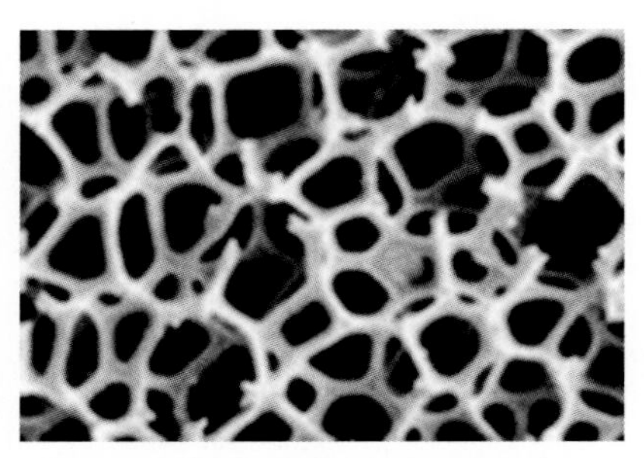

Figure 15.3 *Nickel foam – Incofoam.*

The nickel-cadmium batteries have been superseded in most applications by nickel-metal hydride (NiMH). Cadmium is highly toxic, and the service performance of the metal hydride is generally better. It generates the same voltage as the Ni-Cd and could therefore be used as a direct replacement in existing applications, and it can deliver a higher total amount of power. Early versions had the one drawback that they leaked power when not in use. In NiMH batteries, it is hydrogen that is the source of power. On charging, the input electricity releases hydrogen from the water in the electrolyte. The hydrogen combines with the metal in the negative electrode and the released hydroxyl ions oxidise the nickel hydroxide in the positive electrode, creating the NiO(OH) active mass. The reaction is reversed on discharging as the hydrogen is released from the hydride and the active mass reverts to $Ni(OH)_2$. Since the hydrogen must diffuse from within the hydride alloy there is a limit to the rate at which power can be withdrawn, but in most applications, this is an advantage.

The use of NiMH became widespread in the 1990s; they powered many portable devices. The portable computer became a battleground for achieving ever decreasing weight and by 2000 a higher powered, lighter alternative was available, the lithium ion battery. This soon took over the market, but the characteristics are different from the earlier batteries and new applications had to be designed around them. The nickel-based batteries have not, however, completely yielded to the lithium-based competition.

In 1997 Toyota introduced the first dual-engine car, the Prius. It had a 1.8 litre petrol engine and an electric motor. When the battery is fully charged, it is possible to drive the car without using the petrol engine for about two miles, but the main significance of the design is that the battery pack that powers the electric motor is used to capture energy expended in braking. The design depends on brushless motors that can work both as motors and generators and they switch to generating mode when the car is braked. The electric motor can then contribute energy when more power is needed, making the car much more powerful than a normal 1.8 litre vehicle. Environmentally there is a major benefit in that when at rest, at traffic lights or in queues, the petrol engine stops. There are then no emissions; the vehicle effectively jump-starts, using the electric motor to reignite the petrol engine when the lights change, or the queue progresses. In slow queues the electric engine operates continuously while there is sufficient power in the battery pack. Overall, the advantage of the hybrid design is that it makes cars with petrol engines as fuel efficient as the smelly, dirty, diesel engine cars.

The familiar Prius design is the second generation. It contains a NiMH battery that can supply 1310kWh. Introduced in 2003, it is said that no Prius has had to have the battery pack replaced because of its failure. The smallest of the hybrid cars made by Toyota, the Yaris has a 45kW electric motor and a 0.9kWh NiMH battery. In these cars the fuel consumption (miles per litre) actually recorded (as distinct from the manufacturer's claims) is comparable with that for a diesel engine car of the same general type, but without the high levels of pollution.

Nickel metal hydride batteries do not feature in the all-electric cars that manufacturers are producing, following from widespread legislation to eliminate internal combustion engines from the roads. These have lithium-ion batteries, which also contain substantial amounts of nickel. However, it is a significant advantage of the NiMH battery that it is very safe. That has not been the case with the higher-powered lithium ion variety. Overcharging can cause fires and explosions, as Samsung discovered when they were forced to recall the tablet computer, the 2016 Galaxy Note 7 model, because many of their products had caught fire. Apple's iPads and iPhones have also had problems.

For devices in mobile applications, weight is a major selling point. Nickel-containing batteries made portability possible but were soon considered to be too heavy and have now been overtaken by the lithium-ion family. These gave not only reduced weight but longer operation between charges and higher power, mainly to support bigger displays. The two batteries shown in Figure 15.4 were acquired either side of the change.

The NiMH battery on the left, purchased in 2000, is from an inexpensive Umax laptop. It generated 9.6 volts, delivered 3800 mAh and weighed 500g. The Li-ion battery on the right, from a Compaq of 2002 generated 14.8 volts, delivered 4400 mAh and weighs 425g, enabling it to power more memory and a much bigger screen.

Figure 15.4 *Computer batteries.*

However, many devices are portable only in the sense that they liberate the user from trailing wires around. The notable example is the multiple handset telephone array, where the signal is transferred from the base station by radio and the power to allow it to be used is from a rechargeable battery.

The familiar image in Figure 15.5 shows clearly that in wireless telephones NiMH batteries should not to be superseded. They are cheaper and safer than any likely replacement. It is not advisable to power such units with Li-ion batteries. When not in use, they sit on their charger. Over-charging a NiMH battery would damage the battery, overcharging an Li-ion battery could create a fire hazard. However, ignorance and indiscriminate marketing is resulting in Li-ion batteries being chosen for inappropriate applications. For instance, a commonly available smoke alarm of the type that is attached to mains power, contains a rechargeable Li-ion battery. An NiMH battery would be just as good.

Figure 15.5 *One of three telephone handsets on its charging station.*

Hydrogen is the most important ingredient of these rechargeable batteries and the essence of their behaviour in this application is likely to have wider significance in dealing with energy demand in the future. It will require large scale generation and storage of hydrogen.

Early work on storing hydrogen in metals was carried out at an INCO facility in New Jersey, led by Gary Sandrock.[3] Focussing for obvious reasons on alloys based on nickel, it was found that several of the alloys have very favourable conditions for absorbing and desorbing hydrogen. At room temperature, a lanthanum-nickel intermetallic compound, $LaNi_5$ could be charged with hydrogen at two atmospheres pressure. Lowering the pressure permits the hydrogen to leave the alloy ($LaNi_5H_6$ when fully charged). The hydrogen diffuses out of the alloy on desorbing and cannot burst out explosively. At full charge the hydrogen content is just 1.37% of the mass; that is the hydrogen in one gram-molecule of the hydride is just 6 grams; the metal is 432 grams.

In the hydride, the hydrogen acts as a metallic alloying element. It gives up its electron and sits in the crystal lattice as a proton. In a battery, when the proton is released, it reacts instantly with the water in the electrolyte and the active mass to reduce the $NiOOH$ to $Ni(OH)_2$.

There is much interest in hydrogen absorption in solids as a means of providing a way of storing energy from periods of plenty to use when there is a shortage. Most renewable energy systems, solar cells, and wind or wave turbines do not have the steady output that can be provided by fossil fuel, while it lasts. (Fossil fuel is energy that has been stored over millennia and is not inexhaustible, in fact the most convenient ones will be gone in about two generations). Storage in metal hydrides offers several attractions. More hydrogen can be stored in a piece of metal as a hydride than in a container made of the same volume of a metal of the highest strength. The screening of the proton charge by the electron cloud in the metal lattice allows hydrogen nuclei to get closer to each other than is possible in compressing or liquefying hydrogen gas. The volumetric energy density of hydrogen in $LaNi_5H_6$ is about the same as that of methane. The rare earth-nickel intermetallics are among the best for the absorption and desorption characteristics. However, they will never provide the hydrogen to power cars because of their weight penalty.[4] But they have once again been an important contributor to opening a new area for advancement.

There are applications where the convenience in charging and discharging the hydrogen and the energy density of the $LaNi_5H_6$ has potential, particularly in mobile equipment, where weight is an advantage. Equipment that is used for lifting weights needs to carry counterbalancing ballast. Fork lift trucks are frequently battery powered. They could just as well be hydrogen powered with the hydrogen stored in a hydride, although it is unlikely that hydrogen will supplant the cheap lead acid batteries from this application, even though one hydrogen charge would last for a lot longer. Secondly, most boats require the use of ballast to keep them upright. This is a more promising application, because the interval between charges is important. The hydrogen could either power an internal combustion engine or a turbine or produce electricity directly in a reverse electrolysis process. The latter is preferable, since the effluent is pure water. Internal combustion using any fuel burnt in air produces some nitrogen oxides.

The reverse electrolysis process uses a device called a PEM cell. The acronym stands for Proton Exchange Membrane. The membrane is made of palladium or an alloy that mainly consists of palladium. This can absorb and desorb hydrogen readily. Molecular hydrogen supplied to one side of the membrane splits into protons and electrons on absorption. Continuing absorption results in protons being displaced on the other side, where the environment contains oxygen. Reacting with the oxygen, the electron that the proton has left behind passes around a circuit to complete the formation of water molecules. There are other designs using polymer-electrolytes, but the principle is similar.

There is currently a prominent example of the use of this technology in Birmingham. (Figure 15.6).

The hydriding alloy used in this boat is FeTi. The hydrogen in $FeTiH_2$ can reach 1.89% of the total weight. It is a slightly less easy alloy to charge with hydrogen but is considerably cheaper. Lighter alloys, giving up to 9% of hydrogen by weight are available, but they all need to be heated to a high temperature to desorb the hydrogen. At present, there is no hydrogen storage alloy that can be used to power a people carrier. Current hydrogen-powered vehicles use pressurised gas and need to take up a significant volume to achieve a useful range. In the canal boat application shown, the hydrogen-sourced power from the PEM cells operates a permanent magnet electric motor. The 2.5kg of hydrogen the boat has charged into the metal hydride can take the boat about a hundred miles.

There are many potential applications of hydrogen storage in metals that are not concerned with mobility. In principle, almost any change

Figure 15.6 *The Ross Barlow, a hydrogen powered boat developed by Professor Rex Harris at Birmingham University.*

of state that is reversible can be used in a heat engine.[5] This is most familiar in a refrigerator which uses a readily vaporised liquid to move heat from inside a box and transfer it to the exterior, by absorbing heat when vaporising and giving it up when condensing. This is done on a larger scale with a heat exchanger used to cool or heat space. The absorption and desorption of hydrogen by an alloy can be arranged to perform this function. A schematic diagram of the arrangement is shown in Figure 15.7.

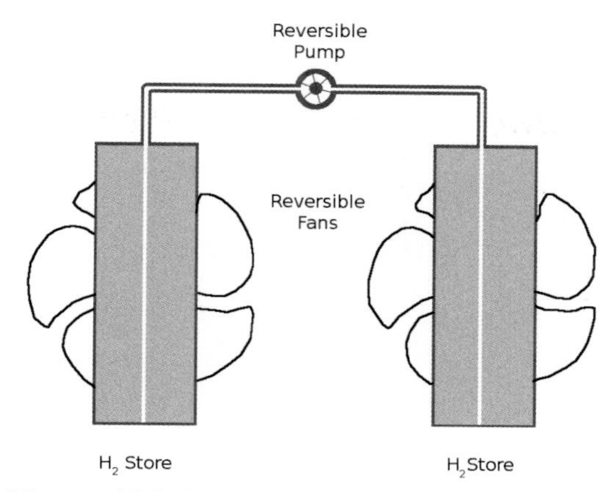

Figure 15.7 *Schematic diagram of a hydride heat pump.*

The evaporation-condensation cycle used in most refrigerators is achieved by moving the liquid and vapour around a circuit, with the movement always in one direction, but this is not possible when the hydrogen must move from one store to the other. The arrangement requires cycling, so that the hydrogen moves back and forth.

The store that is, at any instant, being charged with hydrogen gets hot, because the reaction is exothermic. The discharging store needs heat to be input and so withdraws heat from its surroundings. The schematic diagram is for space heating and requires a reversible pump to pressurise the hydrogen gas towards the absorber and create a partial vacuum in the desorber. It also requires two reversible fans, so that the heat is vectored towards the space to be warmed and from the space to be cooled. There are several designs of working system, see for example reference 5, and other arrangements can be made, possibly using heat pipes, to move heat from one place to another. Such equipment can be used for waste heat recovery.[6]

The pressure-temperature variation of the absorption of hydrogen is shown in Figure 15.9 for two of the best alloys for operating around ambient temperature and pressure.

As the pressure is supplied by the pump to the absorber, its temperature will rise, but the shape of the curves show that it will continue to absorb hydrogen. At the same time the desorber can extract heat from a lower temperature environment because it is subject to a

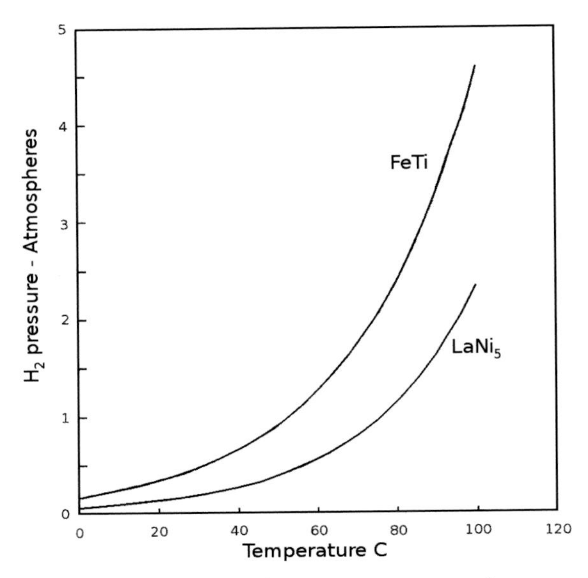

Figure 15.8 *The dependence on temperature and pressure of the absorption of hydrogen into* $LaNi_5$ *and FeTi.*

low pressure. The choice of alloy depends on the temperature and pressure regime in which the equipment would operate. There are many alloys that perform at different levels.

There are many other potential uses for the storage of hydrogen in alloys and $LaNi_5$ is one of the most useful of the available materials. The ability of nickel to release atomic hydrogen is key to its performance in this field, as it is in batteries, and, as we shall see in Chapter 17, in catalysis.

16

Health and Well-Being

The most commonly known fact about nickel is that it can cause contact-dermatitis. This was first recognised by Josef Jadassohn (1863–1936) in 1896. It is experienced when cheap nickel-plated ear-rings or pendants, and metal fittings on brassières give rise to skin irritation. It has led to paranoia about anything that contains nickel and there is much misinformation in supposedly authoritative sources. For instance, the Wikipedia article on the subject claims that sensitisation can be caused by stainless steel.[1] Most commonplace stainless-steel articles contain little or no nickel – see Chapter 8 – and in those that do, the surface is composed of chromic oxide (Cr_2O_3).

Medical advice errs on the side of taking no chances. For instance, the Mayo clinic asserts, without any evidence, that anything containing nickel is likely to cause the allergic reaction.[2] Its claims, that anyone who works in the nickel industry is at risk, is true if the worker is ignorant of the problem and takes no precautions. Certainly, nickel platers were once affected, but it is now so well-known that only the careless are likely to suffer. Nevertheless, there clearly is a problem if unalloyed metallic nickel is worn next to the skin.

There is also a concern that the cheaply produced coinage which uses nickel plated steel as monetary tokens, is a hazard for people who constantly handle money but there is little evidence of this so far. Bank tellers and even shop keepers now use automated machines to count coins, so the level of contact is less than it would once have been.

The accounts of the effect of nickel in contact with the skin are generally based on an assumption that those affected are allergic to nickel. It is, however, possible that the effect depends on how copiously an individual perspires. In chapter 11 it has been noted that nickel-containing alloys are susceptible to failure in a chloride-containing environment. A possible cause of this is that nickel can take hydrogen from water molecules. A perspiring body is a chlorine-containing environment and, as in the case of stress-corrosion cracking in stainless steels, it is possible that the presence of nickel results in the build-up of hydrochloric acid, which will certainly be an irritant on the skin. Copious perspiration might be expected to ensure that the acid is very dilute, perhaps less sweaty people are more susceptible.

With this background it is surprising that an alloy containing about 50 atomic percent of nickel is used increasingly in medical and related areas. The alloy, an intermetallic compound, is NiTi. It can be processed to have the property that it remembers its original shape, so that if it is bent out of shape it will revert to that shape on the application of a modest level of heat. The range of applications where this is a benefit is not limited to the medical. For instance, compositions can be made that will flip-flop between two different shapes depending on the temperature. This has been used to make self-activating hinges for green-houses, to permit ventilation when the temperature is high. A further potential application is in delivering mechanical work by the application of heat. As seen in the previous chapter, any reversible change of state or composition can be used to design heat engines.

The most notable medical application is in stents. The use of a shape memory alloy can greatly simplify the operation to insert the stent. This is a hollow cylinder that is inserted to keep arteries open to prevent heart attacks. Making the stent of NiTi, it can be rolled up to form a small cross-sectional area and attached to the catheter that is used to reach the heart through an artery, probably starting in the leg, a long way from the heart. When in position the body-heat causes the rolled-up shape to unwind and form the tubular shape that is to keep the artery open.

In this application, the nickel alloy is to be in contact with the tissue of the patient so long as he or she survives. Evidently the medical profession has no concern about the likelihood of an allergic reaction. The NiTi has its surface covered with TiO_2, but the environment in which it operates is also chlorine-free and contains a copious flowing liquid.

Opticians also do not worry that the nickel in NiTi will cause problems. Among the most expensive frames that can be had for spectacles are those made of the alloy. The arms of the frames are intended to be in long term contact with the ears and the side of the head. The advantage offered by the NiTi is that if the frame is bent in any way the application of a low level of heat will provide an immediate repair.

Perhaps the last place that anyone concerned about an allergic reaction to nickel would wish to put one of its alloys is in the mouth. Yet this is precisely what is done by dentists who use NiTi to make teeth braces. Rather ugly braces, designed to allow progressive adjustment of the positions of individual teeth, have screws that can be adjusted as the teeth move to the desired positions. Neater braces can be designed if the brace is auto-reactive. This can be achieved by making the brace in NiTi. In the heat generated in the mouth, the brace, which is made initially in the form that is the ideal configuration for the teeth, is bent to fit the initial condition. When placed in position in the mouth it will seek continuously to return to its initial condition.

In the case of the use of NiTi for spectacle frames, the application probably depends on the titanium dioxide coating remaining intact. When the application is inside the body the chemistry is very different from that of the skin. The greater volumes of liquid do not permit highly acidic conditions to develop and there is generally an absence of chloride ions, which appear to be the most significant hazard for nickel-containing materials.

How is it that NiTi can be used and perform in the manner described?

The absence of any reaction to the material is because, like stainless steel, it is covered with a layer that does not contain nickel and which will restore itself if breached. In the case of NiTi, the layer is TiO_2. Titanium dioxide is bio-compatible. Titanium alloy is used for many of the prosthetic devices for hips and knees and for pinning bones. It is only a little over half as heavy as steel but capable of similar strength and is more compatible with body tissue than the cheaper stainless-steel alternative.

The shape-memory effect is not unique to NiTi. It was first noted in 1932 in a gold-cadmium intermetallic.[3] Like NiTi, AuCd has an ordered body centred cubic structure, Figure 16.1.

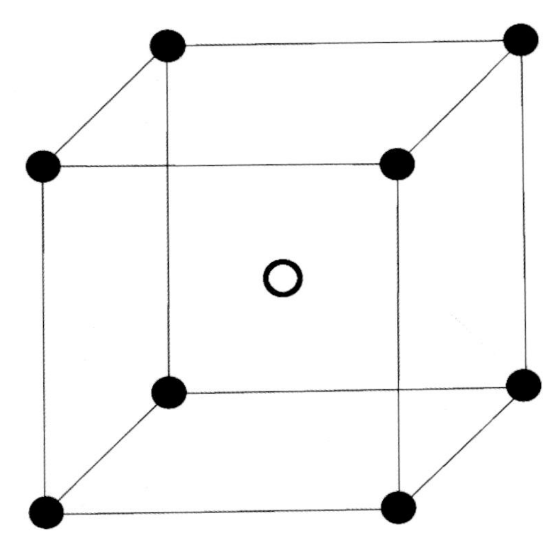

Figure 16.1 *The ordered body centred cubic structure, common to many equiatomic intermetallic alloys. The eight cornering atoms are shared between eight of the cubes; the structure could be drawn by interchanging the two species.*

Also, like NiTi, it undergoes a martensitic transformation at low temperatures. In the case of AuCd, it is to an orthorhombic structure. That means that the crystal structure is brick-shaped, with its three edges of different lengths. In NiTi the martensite is similar but slightly off rectangular; small amounts of other constituents, such as copper, can modify the structure to be like that of AuCd. The orthorhombic structure is shown in Figure 16.2.

The cubic structure of the ordered body centred cubic high temperature structure is outlined to show the relationship to the orthorhombic structure. There is little difference between the two atomic arrangements, just a slight shift in the relative position of the two atomic species. In the modified form found in NiTi, the

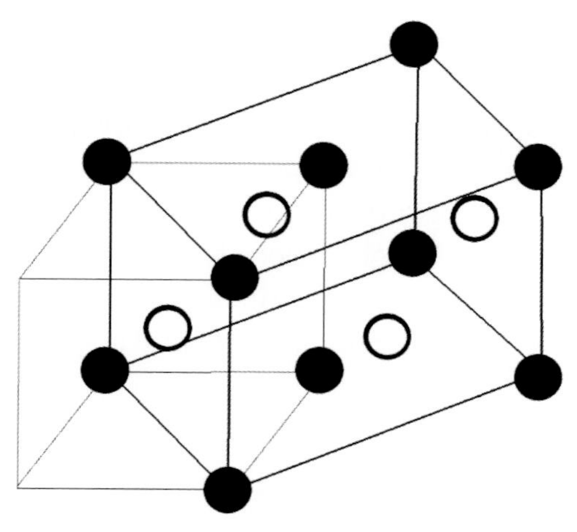

Figure 16.2 *The Orthorhombic structure of AuCd and some NiTi alloys.* ● *represents Ti atoms,* ○ *the Ni atoms.*

orthorhombic structure is distorted slightly as if it were pulled along a body diagonal.

As previously discussed, martensitic transformations occur without diffusion. They result in local shape changes. If, as in the case of maraging steels, there is little volume change and the shape changes are randomly distributed, they may cause no overall shape change. However, if the transformation to martensite occurs under stress, the transformations that are favoured by the applied forces will contribute to an overall change of shape. This is called transformation induced plasticity. The application of stress alters the stability of the initial structure, so it is a condition of this type of behaviour that the two structures are almost equally stable. When the stress is applied the martensite is preferred; when there is no stress, the ordered bcc phase is favoured. However, once the martensite is formed, there needs to be some energy supplied to cause the structure to revert. This is usually supplied by modest heating. The formation of martensite creates internal stresses in the material, and when the thermal activation causes reversion of the structure, the internal stresses are unloaded by reverting to the original form.

In alloys that are finely balanced in terms of the stability of the two phases, the material can be artificially forced to cycle back and forth along a particular shape transformation to create a pattern of residual stresses that work in both directions. This then allows items, such as the active hinge for greenhouse ventilation, to flip flop-from one position to the other depending on the temperature.

Many applications have been found for this behaviour and there will no doubt be more, all highly specialised and of small volume, but the nickel-containing alloys are the universal choice for them, mainly because the general properties of strength and corrosion resistance are always required.

17

Catalysis

Nickel is one of the three most commonly used catalysts. All three are in the same column in the periodic table (Figure 3.6), indicating that they have many chemical similarities. Nickel is more reactive than palladium, which is more reactive than platinum, but none are highly reactive. The essence of a catalyst is that it assists reactions to occur but does not participate irreversibly in them. Nickel is already widely used as a catalyst but there is much effort devoted to trying to develop its properties so that it can replace both palladium and platinum. Nickel, the 23rd most abundant element in the earth's crust, at 80 parts per billion, is cheaper and more readily available than either palladium (0.6 parts per billion) or platinum (1 part per billion),[1] and, furthermore is a constituent of the earth's core so that the amount available is likely to be replenished by volcanic activity.

Paul Sabatier (1854–1941) first observed the catalytic effect of nickel in hydrogenating organic compounds, when he directed a stream of ethylene (C_2H_4) at nickel (and cobalt or freshly reduced iron) held at 300°C.[2] The metal became coated with carbon, but the gas had become ethane (C_2H_6). There being no excess hydrogen available, the metals catalysed the formation of the more stable ethane from the atoms that were available. Introducing hydrogen into the mixture resulted in all the carbon being used in the resultant ethane.

These experiments were carried out in 1897. Sabatier later showed a similar effect with acetylene (C_2H_2). Other metals could be used to catalyse these reactions, but none that he tried were as vigorous as nickel. Sabatier shared the Nobel Prize for Chemistry with V. Grignard in 1913.

Since these pioneering studies, nickel has been used in all applications of organic chemistry, in petrochemical, pharmaceutical, and food industries, to modify molecules by hydrogenation. It is one of the most common transformations used in industry and affects all our lives. Those who aim to eat healthily will be aware of hydrogenated natural fats, which may not be good for health, but there are many other reactions that involve the hydrogen modification of molecules.

One of the most widely used catalysts was first used for hydrogenating fats. Raney nickel was developed by Murray Raney (1885–1966) in 1925.[3] It is a finely divided nickel powder in which all the particles are porous giving a large surface area. It is made by immersing nickel in liquid aluminium, which produces nickel aluminides. Alkali-leaching removes most of the aluminium to create the porous structure. A wide range of organic reactions can be catalysed with Raney nickel and it is used in removing sulphur from organic molecules.

Palladium and platinum are generally better catalysts than nickel, but the search for methods of improving nickel's performance in some of the applications where these two metals are currently used is of great urgency. Emsley estimates that the world reserves of platinum in 2005 were 30,000 tons.[1] The best-known use as a catalyst is in car exhaust systems. Although only one or two grams are used per car, or 12 to 15 grams per lorry, only 25% of this is recovered by recycling. Thus, one of our rarest materials is being spread in a monolayer across the surfaces of the world's roads. Unfortunately, nickel is not a good replacement in this application. It would capture any carbon monoxide and the nickel carbonyl emitted into the atmosphere would be a greater hazard than the NO_x and petrochemical fumes. However, replacing platinum elsewhere would be a benefit.

Palladium, which is even rarer than platinum, is currently necessary in PEM cells (see Chapter 14). Any hopes of making hydrogen powered vehicles widely available will depend on finding an alternative material, as well as developing a lighter hydrogen storage system.

The way in which catalysis works is extremely complicated with many and various electronic states being available to break and make bonds in the compounds that are being modified.

It is a feature of the transition metals that they can adopt different valence states. In the early period elements, this makes the metals highly reactive so that they are more likely to participate in a reaction than catalyse it. The more noble metals, although inert in reaction terms, can adopt different electronic states at their surfaces. It is not a coincidence that the metals that most readily absorb hydrogen are the ones most used as catalysts.

To absorb hydrogen into the metal lattice the H_2 molecule must split into its atomic parts. As Sabatier points out,[2] hydrogenation reactions had been observed before he discovered the catalytic effect of nickel, but it was necessary to generate nascent hydrogen to make them happen. In the case of the use of nickel in this, its most common catalytic role, it appears that the hydrogen molecule on reaching the nickel surface and dividing into two atoms is instantly available to react with a hydrocarbon that can achieve a lower energy state by reacting with the atomic hydrogen.

From Sabatier's experiments it is evident that the nickel generates atomic hydrogen and it is the hydrogen that breaks the carbon bonds. In the case of the conversion of ethylene into ethane, as shown in Figure 17.1, the double carbon bond is broken to allow the attachment of two further hydrogen atoms. In the case of the conversion of acetylene the breaking of the triple bond gives four further sites for attaching hydrogen atoms.

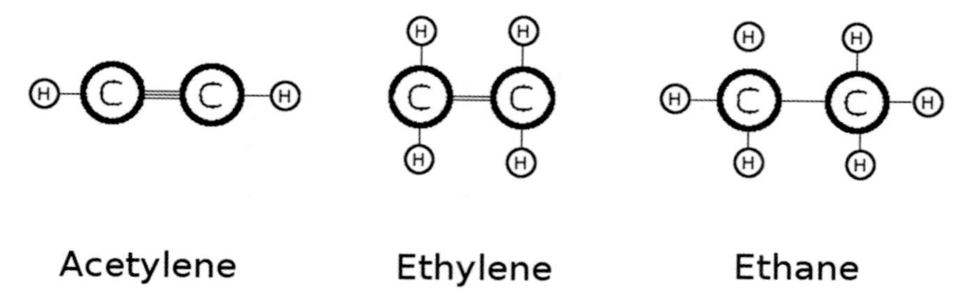

Figure 17.1. *Carbon-hydrogen compounds.*

If additional atomic hydrogen is supplied the reaction:

$$C_2H_4 + 2H = C_2H_6$$

is known to occur without a catalyst.[2] But when there is no additional hydrogen, the C-H bonds of the ethylene must be cut to liberate the atomic hydrogen. The reaction that causes the breaking of the bonds involves the electronic structure at the surface of the catalyst. One ethylene molecule supplies the hydrogen for two ethane molecules, leaving two carbon atoms each time. This reduces the gas volume by a third. In the acetylene reaction, the gas volume is reduced to a third.

The low reactivity of nickel with oxygen, nitrogen and carbon indicates that the molecules of these species are not separated into atoms at the interface with nickel as the hydrogen molecules are. The size of the atoms is the major factor. Rather than separating carbon from oxygen, when the two are combined as carbon monoxide, four molecules of the gas pick up one nickel atom.

Metallic nickel was where the effectiveness of the metal in catalysis was first noted, but the search for more effective variants has created a branch of chemistry of ever increasing complexity, using organo-metallic compounds based on nickel to catalyse many different reactions involving a variety of radicals attached to organic arrays. The developments are particularly important in the field of medicine, where the many compounds that are prescribed and are made in modest volumes involve a wide variety of reactions in their manufacture. Some background to these activities can be found in Tasker et al.[4]

18

Forward

Every element is unique. The special uniqueness of nickel lies in the number and scope of its properties and its position in nature.

Isotopes of iron and nickel are the most stable conglomerations of protons, electrons and neutrons in the Universe. Fe_{58} and Ni_{62} are most stable isotopes and it is for this reason that the cores of planets are commonly composed of these elements and it explains why heavier isotopes are less common.

That nickel is near to the end of the first long period in the table of elements (Figure 3.6) puts it in the company of the noble metals, as the first and lightest of the elements that have good corrosion resistance and active surface properties. It also puts it close to iron and cobalt and in the group of magnetic metals.

It forms strong interatomic bonds, giving it a relatively high melting temperature and enabling it to alloy freely with many other, especially the most common, metals.

The discovery of these attributes followed directly from the separation of the element to make a pure source available, which was due to Charles Askin and Edward White Benson in Birmingham. The metals became available at just the right time to be taken up by the many people who were interested in exploring the behaviour of materials.

Faraday, Percy, Chevenard and Hopkinson made alloys and explored their strength and magnetism, quickly developing the major applications that started to appear early in the twentieth century.

In time to meet this demand, Ludwig Mond elaborated his idea of how better to purify the metal at a laboratory on Birmingham Heath and brought a large source of the metal into play.

In the first half of the twentieth century most technical progress could not have occurred without the contribution made by nickel. Wars stimulate progress. The first major war revealed, as never before, the importance of nickel in making steels strong and tough. The second war gave us radar and computers and, most significantly, gave rise to the interest in gas-turbine powered aircraft.

In several instances the initial function has been overtaken by later developments, but the role of nickel in helping to provide a stepping stone to a better future was significant.

The AlNiCo magnets that made radar possible were overtaken by other methods, and other high-power magnetic materials have been developed, but when there was a need for radar, AlNiCo did the job. In computing, the contribution of nickel was not in the front line, but, without the reliable thermionic values that used nichrome heating elements, the code breaking at Bletchley Park would have been a frustrating and ultimately pointless exercise.

The thermionic valve was the first semi-conductor, but today the word has been adopted as relating only to silicon or other solid-state materials. These came too late to help the code breakers.

AlNiCo magnets are not now used in brushless motors – one of the major applications of high-powered magnets, used especially in wind turbines – but they still command significant markets, where the power available is sufficient and especially when there is a need to perform at elevated temperatures. Their Curie temperature is around 860°C, compared to 400°C for the NdFeB magnets. Samarium-cobalt magnets can operate at up to 800°C, are about three times as strong magnetically as the Alnico but are brittle and expensive.

There are many instances where the functionality of nickel is not obvious but without it many items we regard as essential would not work.

The switching of kettles, toasters, ovens and heating systems depend on small amounts of the low expansion alloys that use nickel and there are many other hidden uses in electrical circuits and small transformers that require its presence. But integrated circuity using silicon chips has replaced many of the early uses.

The use of nickel has, nevertheless, benefitted from miniaturisation. Small systems need small amounts of power and are required to be portable. The little batteries used in cell phones and tablet computers all make use of nickel. Some, less portable, devices use nickel salts as the active ingredient of the battery, while others, like the button cells used in key fobs, hearing aids and remote controllers are generally in nickel-plated cases.

Although in a few cases nickel has helped to advance particular technologies but has been superseded by better methods, in the applications where most of the metal is used, there is no likelihood that it will ever be replaced.

The 16% used annually in plating artefacts may change slowly as fashions change and will be affected significantly if cheap plastic goods lose their appeal, but many applications – the battery cases already mentioned, the inner lining of electric kettles and many others – are unlikely to be discontinued. But for quality chromium-plated goods, an initial coating of nickel will still be needed.

It would be possible to replace stainless steel as the cladding for buildings and white goods with titanium sheet. One of the Liverpool Cathedrals has a titanium clad roof, but the price of titanium is held at a high level by an industry that restricts its production and promotes it as being highly valuable, even though it is the fourth most common of the engineering metals and is extremely easy to process to semi-finished products. Its abundance is evidenced by the fact that most paints are now based on its oxide. Unless a new entrant into the titanium industry develops an improved method of extracting the metal from the readily available oxide, it is unlikely to challenge stainless steel as a cladding material. It is, however, for its high temperature capability that nickel is unchallenged. Many have tried to replace nickel alloys in withstanding loads and corrosion at high temperatures. All of them have been found wanting. Ceramics can be stronger at high temperature, but they must also resist failure at low temperatures, which they are unable to do. Intermetallic compounds have also been considered but are equally limited, except as thermal barrier coatings.

Many workers around the world have contributed to the development of the alloys that are essential to the operation of high-performance gas turbines, but none have contributed more than the research laboratories of INCO. The journey to these advanced materials started, as did other important innovations, in Wiggin Street, Birmingham.

Metallurgical research had its heyday in the first half of the twentieth century. Every area of human discovery is like a mine or an oil well; it becomes exhausted over time. That the research activity and the association with the development of the nickel industry is no longer evident in Birmingham is not a matter for regret, but that it started there, and contributed so substantially to material progress, is to be celebrated and remembered as an important part of the culture of the area.

Nothing remains in Birmingham to show that the nickel industry started there. The site of the first office occupied by Evans and Askin is now occupied by a bank. Wiggin Street has one mill building remaining. It houses a small company making metal components for the auto-mobile industry. But equally there is only one brass mill in the whole of the West Midlands and the steel mills are long gone, mostly converted to offer the retail experiences that drive our economy.

It remains true that some of the most significant developments that shape our lives today started here. Taylor and Boulton had increased productivity long before Adam Smith took the credit for it by articulating the principle of the division of labour. Boulton and Watt had begun the process of freeing production from the limitation of the available water power, and some of the most significant developments in the understanding and availability of material capabilities were developed in and around Birmingham. Cast iron, much overlooked, but an essential ingredient of industrial progress, was improved and characterised, but pre-eminent among the developments that have made changes to the lives of everyone are those that have followed from the first development of a nickel industry.

Appendix

1. Atoms and their behaviour

1.1 Atomic Structure

Explaining the order of the elements in the Periodic Table is the starting point for Quantum Theory. The inert gases are critical in understanding it. It was recognised that it is not atomic weight that differentiates the elements, but the number of protons in their core; that is the atomic number. There is a matching number of electrons in orbit around the nucleus. The nucleus can contain neutrons; different numbers of neutrons are possible, giving rise to isotopes of the elements. The atomic numbers of the inert gases are 2, 10, 18, 36, 54, 86 respectively. Evidently these represent stable arrays of electrons, so helium with 2 electrons has no capacity for forming bonds. A further 8 electrons result in another stable configuration and so on. The electron configurations are given below.

																	Atomic Number
Helium	2																2
Neon	2	2	6														10
Argon	2	2	6	2	6												18
Krypton	2	2	6	2	6	10	2										36
Xenon	2	2	6	2	6	10	2	6	10	2	6						54
Radon	2	2	6	2	6	10	2	6	10	2	6	10	2	6	10	14	86
	1s	2s	2p	3s	3p	3d	4s	4p	4d	5s	5p	5d	6s	6p	6d	6f	

The sequence is rationalised by assigning to each atom a series of electrons that are characterised by four numbers; three are integers, the fourth, the spin, can be either $+\frac{1}{2}$ or $-\frac{1}{2}$. The integers, **n**, **l**, and **m** take values in the ranges, $0 < \mathbf{n} < 8$, $0 < \mathbf{l} < (\mathbf{n} - 1)$, while **m** can take any integer value between **-l** and **+l**. All permutations of the three variables correspond to permitted energy states. **n** corresponds to the row number in the periodic table. Bearing in mind that each combination of **n**, **l**, and **m** has two spin values, it will easily be calculated that the available combinations generate the sequence of Figure 3.6. Although the sequence may appear to be purely numerical, it has the virtue that it explains many of the similarities and behaviours of the various elements.

1.2 Magnetism

Magnetism is one of the most readily recognised of properties, but also one of the least easily understood.

Some factors can be pointed to. As discussed above, each electron attached to an atom is defined by four quantum numbers. Three define its total energy, the fourth is the electron spin. This can take only two values + ½ and - ½. It can be recognised from Faraday's earliest experiments with electricity that a changing current generates a magnetic field.[1] When electron pairs have opposite spins, the fields they generate cancel out, but if spins of one direction predominate their fields add up to make the material magnetic. The more difficult matter to explain is why the ferromagnetic materials choose to prefer one spin over the other, which they do. It is not the case that the spins become aligned when a magnetic field is applied. Individual atoms always, at temperatures below the Curie point in these materials, prefer one direction of spin for one of the highest energy electrons.

In the demagnetised condition, atoms cluster in domains, each of which contains atoms with their spins parallel, but not coordinated between domains. Magnetising the material causes the domains to arrange with their spins aligned. This happens by a mechanism similar to grain growth. Increasing the field causes the domain boundaries to move, one direction of spin gaining by reducing the volume of the other. As in grain growth, the boundaries can be pinned by second phase particles, making it harder to magnetise or demagnetise the material.

In each domain the spin is aligned with a preferred direction in the crystal. It follows that if all of the crystals have this direction aligned, the magnetic strength will be maximised.

1.3 The Iron-Nickel Anomaly

The behaviour exploited by Guillaume in developing Invar is based on the behaviour of mixtures of nickel and iron atoms.[2]

It is evident from Figures 3.9 and 8.2 that iron is a stronger magnet with a higher temperature at which the magnetism disappears than nickel. If the law of mixtures applied, the magnetism would decrease gradually from iron to nickel but, since the system has more than one phase, other effects take over.

As pointed out in Section 2.4 below, a nickel atom is slightly larger than an iron atom in ferrite, but the reverse is true in the fcc phase. The difference is small, and this has the important effect that inter-diffusion is slow. If a true equilibrium could be reached, a 36% nickel alloy would consist of 56% of an ordered compound, Fe_3Ni and 44% of another compound, FeNi. Both are ordered, and based on the fcc structure, although FeNi is probably slightly tetragonal, because the two elements lie in alternate layers. In practice, this does not happen; the elements are, instead, distributed among each other fairly randomly. It is likely that in small regions where the iron atoms are more concentrated than on average, the nickel atoms will behave more like those in ferrite, assuming a smaller size than the iron atoms. When the ion size is reduced, it is because the d-electron that confers the magnetism through its spin has joined the valence band, thus losing its magnetic effect. It appears that the state of ionisation at 36% nickel is marginal and that as the temperature rises, the thermal activation promotes the most energetic d-electron into the valence band, resulting in more nickel atoms adopting the smaller size that they adopt in ferrite.

Such an explanation will not satisfy mathematical physicists but is intended to give those who are less expert some idea of the realm in which an explanation lies, in a way that is understandable intuitively. More rigorous attempts at an explanation suggest that the phenomenon is due to competition between two states of magnetism, ferromagnetism and paramagnetism and calculate the variation in electron spin directions as a function of composition and temperature.

1.4 Hard Magnets

The development of the AlNiCo magnetic materials began when Mishima discovered that the addition of aluminium restored magnetic properties to the iron-nickel composition with around 36% nickel (Figure 8.2).[3] Mishima's discovery stimulated Bradley and co-workers at Manchester University to study the phase composition of iron-nickel-aluminium alloys.[4] The results of their X-ray studies of the phases present at room temperature are shown in Figure A.1.

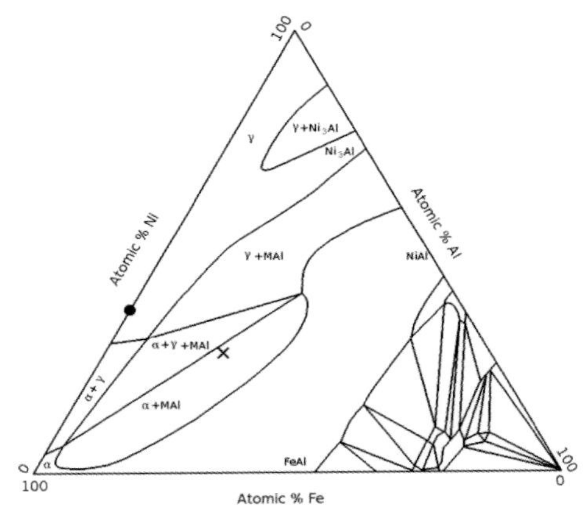

Figure A.1 *Room temperature section of the iron-nickel-aluminium phase diagram, after Bradley & Taylor.*[10]

The circle marks the Invar composition first recognised by Guillaume, the cross is the first composition used by Mishima. The addition of aluminium of 10% by weight is 19% of the atoms present, because aluminium has an atomic weight a little less than half of that of iron and nickel, so the composition selected by Mishima was, in atomic content, significantly removed from the Invar composition. As can be seen, Mishima's alloy is not, like Invar, austenitic; it is a mixture of ferrite and an intermetallic compound which has 50% of its atoms aluminium and the rest a mixture of iron and nickel (shown as MAl, the M standing for either iron or nickel).

The intermetallic phase forms by an ordering reaction at quite low temperatures, so its effect is to precipitation harden the alloy and lock in the matrix structure. The ferrite phase in Mishima's alloy is magnetic, and, when the M is predominantly iron, the MAl phase is also.

The first to anneal an AlNiCo alloy in a magnetic field were Oliver and Sheddan.[5] They reported a 20% increase in saturation magnetisation in the field direction. Annealing in a magnetic field has the effect of making the ferrite slightly tetragonal; precipitation occurs in a manner that locks in the domain structure that the magnetic field creates, making it harder to demagnetise the alloy. A further increase is achieved by controlled solidification; this is the result of magnetic anisotropy. In the bcc structure the direction of easiest magnetisation is along the cube edge. Aligning most of the crystals so that they share a common cube direction enables the greatest magnetic strength to be locked in.

A common method of achieving strong magnetic anisotropy, is by atomising the alloy to produce a finely divided powder and consolidating it in a magnetic field. Before the powder particles stick together in sintering, they have some freedom to align with the magnetic field. This method is used for the neodymium-iron-boron alloys, which can be easily fragmented. It explains much of the reason for the superior properties that this material generates.

2. Some Basic Metallurgy

2.1 Strengthening mechanisms

The mechanisms by which nickel contributes to strength in its alloys with other metals are not, in principle, specific to nickel, nor to the commonest alloys, which are various steels. As noted above, only grain refinement enhances both strength and toughness. Other mechanisms can affect strength but invariably reduce either toughness in body centred cubic metals, or ductility in face centred cubic metals. Metals with a hexagonal structure generally have another problem that is not relevant here that gives them poor ductility. They do not have enough ways of deforming to enable arbitrary shape changes to be achieved.

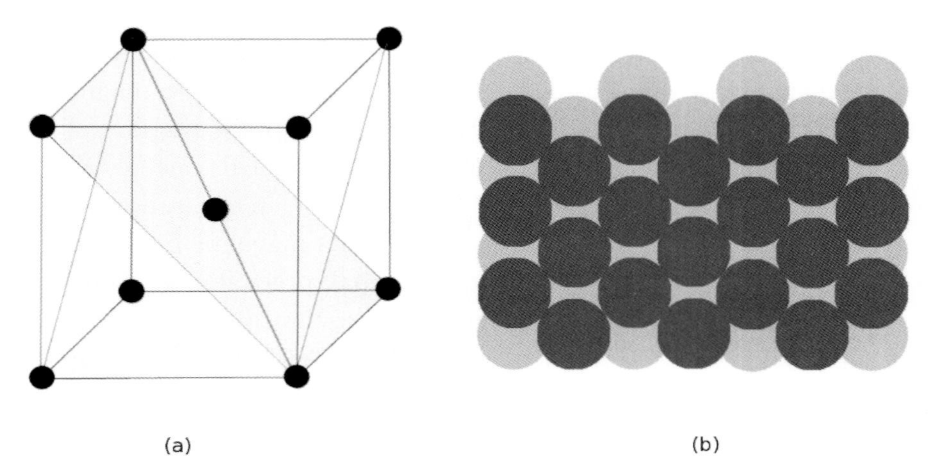

Figure A.2 *Body centred cubic structure showing a) two planes of closest packing and the close-packed direction at their intersection, and b) the stacking of the close-packed planes, each having two close packed directions.*

The dominant means by which metals deform is a process of crystal planes slipping over each other. In the bcc structure, illustrated in Figure A.2, the most widely spaced planes are the body diagonal planes, two of which are illustrated in Figure A.2a. There are six such planes, each of which contain two close-packed directions, which are, as shown in the diagram, the lines common to pairs of the diagonal planes. Looking down on the plane on which slip can occur, in Figure A.2b, the two close packed directions are evident.

Slip does not occur by the planes sliding over each other like a deck of cards but by the movement of defects in the crystals, dislocations. Deformation might be envisaged by thinking of how a worm moves. The rings on the surface of the worm are seen to be compressed together in a small section, which propagates along the length of the worm. In the case of a deforming crystal the simplest analogy for the worm's rings is an extra half plane perpendicular to both the slip plan and the slip direction. This is an edge dislocation, illustrated in Figure A.3, where the lines represent atomic planes, seen on edge. However, the lattice defect that this represents is not restricted to lying at any particular angle to the slip direction. (Those interested in learning more should find a copy of W T Read's book 'Dislocations in Crystals'[6] or A

H Cottrell's 'Dislocations and Plastic Flow in Crystals'[7]). Dislocations are, as can be seen from Figure A.3, the boundary between a region that has already deformed and the yet undeformed volume.

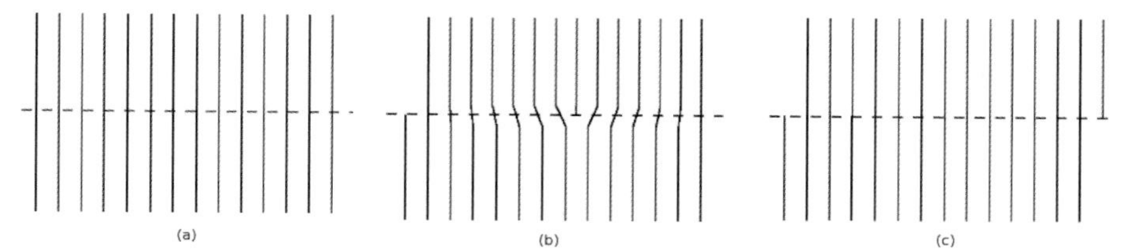

Figure A.3 *Illustrating how deformation typically occurs in crystals. In (a) the crystal is undeformed, in (b) a dislocation has progressed half way across the field of view and in (c) it has run out of the field.*

The process of slip is the same regardless of how the material is strengthened. (There are other, less important, mechanisms that are not usually relevant in nickel containing alloys, although deformation associated with phase changes is mentioned in Chapter 16). The strengthening mechanisms are essentially methods of making slip more difficult. There are four methods that commonly contribute to a material's strength. These are work hardening, grain refinement, solid solution strengthening and precipitation, or second phase particle, hardening.

2.2 Work Hardening

Any metal deformed at low temperatures becomes harder. The stress required to cause deformation rises progressively until the metal fails. Thus, using work-hardening inevitably takes up some of the ductility of the material and makes it more likely to break. The hardening arises because the process of deformation results in crystal defects being left behind as different slip systems interact with each other. The defects make the crystals less perfect and harder to deform.

The simple shear mechanism illustrated in Figure A.3 is the basic mechanism, but it must be recognised that in the face and body centred cubic crystals there are twelve geometrically similar configurations of slip plane and direction; at least five of which are required to operate to impose an arbitrary shape change. These interact with each other and the interactions have many possibilities for adding to the dislocation debris that results from deformation.

2.3 The effect of grain size

Grain refinement affects strength because each crystal must find its own way to respond to the imposed loads. One that is oriented favourably to contribute to the imposed strain may start to deform at a relatively low load, but it then has to find a way of propagating the deformation into neighbouring grains that are not so favourably oriented. A large grain can push harder on its neighbours than a small grain, so that the external force does not have to be raised as much to generate deformation through all of the grains.

For metals that fail by brittle fracture, there is a preferred plane on which the crack propagates. In bcc metals it is a cube face. A crack initiated in one grain is generally unlikely to find that the grains ahead of the crack offer a cube plane just oriented so that the crack can run straight through; it has to deviate requiring a higher stress in order to continue. The stress at the tip of a crack in an elastic solid was shown by A.A. Griffith (1893–1963) to be inversely proportional to the square root of the crack length.[8] Thus, if the crack has crossed the first grain, its length will be the grain diameter and the relationship of the brittle fracture stress to the grain size, as shown in Figure 7.7, is explained.

Quenching iron alloys to produce martensite and subsequently tempering is the most effective way of producing a fine-grained ferrite. It is this aspect of the structure that provides the good toughness. There is also carbide, or other second phase, precipitation that gives additional strength, as described below.

2.4 Solid Solution Strengthening

A solid solution is one in which atoms of the parent metal are replaced by atoms of a different element, without changing the crystal structure. However, they will generally be either larger or smaller, it doesn't matter much which; for a given percentage difference the effect will be about the same either way. If it is imagined that the dark atoms in Figure A.2b are sliding over the grey ones in one of the close packed directions, introducing atoms of a different size produces either bumps or depressions in the two surfaces resulting in a greater resistance to deformation.

This is a very simplistic explanation of the effect. More complex interactions can occur because individual solute atoms may have a strong binding energy with dislocations, occasioned both by size, the region around the dislocation being distorted, and by electronic effects; a dislocation also perturbs the local electron density.

Solution strengthening is always a modest effect. If the size difference of the two species is large, there may be a larger strengthening effect per percentage of solute in the alloy, but the extent of the solid solution will be restricted because of the incompatibility in size.

In the case of nickel in iron, the nickel atom is slightly larger than the iron atom in ferrite, but slightly smaller in the fcc structure. In both cases, there is a modest hardening effect at similar grain sizes.

2.5 Particle Hardening

The term particle hardening is preferable to precipitation hardening because particles of a second phase can be introduced in several ways. They can be introduced physically by mixing constituent particles; they might form by separation during a phase transformation, or, most commonly, they can form by precipitation from solid solution.

Second phase particles come in all shapes and sizes and are located either within grains or at grain boundaries. For low temperature properties, particles at grain boundaries are most likely to have a damaging effect. They will often be flattened to lie in the grain boundaries and most commonly they weaken the material. At high temperatures they can be useful in resisting grains from sliding past each other along the boundaries.

The particles cause hardening in one of two ways. Either the deformation has to by-pass the particles, or the particles must deform with the matrix.

The initial stress needed to by-pass the particles is inversely proportional to the particle spacing, so that for a given volume fraction of the second phase the finer the particles the greater their contribution to strength. As the material continues to deform, the region containing the particle, having not deformed, is surrounded by a region that grows as deformation proceeds containing defects – dislocations – that form the boundaries between regions that have and have not deformed. This is a form of enhanced work hardening, so that as well as starting to deform at a higher stress, the material hardens more rapidly as deformation continues, resulting in lower ductility.

When the second phase particles are separated from the matrix by a clearly defined, localised boundary, the hardening is simply as explained above. When the alloy is one in which the particles are produced by precipitation from solid solution, the hardening may go through a number of phases before a fully separate phase has formed. The precipitation occurs over time during ageing. It starts by the minority atoms clustering together, or in an orderly sequence with the parent atoms. If the atomic sizes are significantly different, this produces lattice distortion, which starts to make it more difficult to move the dislocations. The best-known example is in the Al-Cu alloys. At this stage the clusters will behave as part of the parent structure and if cut by dislocations there will be a small drag effect, because atoms which have come together to reduce the energy of the total system are being pulled apart.

As ageing proceeds the arrangement of atoms will approach the structure of the second phase, but while still quite small may retain coherence with the lattice. They will be associated with greater lattice distortion and be progressively harder to drive dislocations through. This is when the greatest hardening effect occurs.

Further growth of the clusters results in the phase becoming separate, with a clearly defined boundary between it and the matrix and the strength begins to fall. The morphology of the precipitate determines whether the strength in this over-aged condition is useful or not. Plate-like precipitates are more likely to reduce ductility than spherical ones.

The best precipitate systems that have been discovered are in nickel alloys, in which solute clusters remain coherent with the parent lattice and are very stable, while conferring remarkable strengthening effects because of the difficulty of moving dislocations through the clusters. This form of hardening is described below.

2.6 The effect of nickel on the mechanical properties of iron alloys

The role of nickel in all the ferrous materials is to refine the grains size and increase the toughness. How this works depends on some simple effects that will now be described.

When Riley, in his paper in 1889, argued strongly for the benefits of adding nickel to steels, he had only the slightest understanding of how it brought about these benefits.[9] The way in which the materials were processed was based on the practices that had been developed over centuries in working with iron containing carbon. It was known that wrought iron, which has a low carbon content, could not be made harder by heating it to red heat and dropping it into water, but that cemented iron became substantially harder, and more brittle, if reheated and quenched in water. This was the effect of the higher carbon content that had been introduced by heating the wrought iron in charcoal. Then reheating to a lower temperature – called tempering – the hardened steel could be brought to a good combination of strength and resistance to fracture.

Similar treatments applied to nickel-containing steels, containing carbon, yielded significant improvements in both strength and toughness. There are several factors contributing to the effect.

When pure iron is first melted and then allowed to cool down, it solidifies with a crystal structure that is body centred cubic (bcc), illustrated in Figures 3.3b and A.2. Between 1394°C and 912°C it is face centred cubic (Figure 3.3a), reverting at lower temperatures to the bcc form. The transformations between the two types of cubic structure are key to the mechanical properties of iron and its alloys. Most of the important ones contain carbon and it is the form in which the carbon occurs in the alloy and the way in which this can be changed by heating and cooling that accounts for the properties developed.

The carbon is either dispersed as individual atoms in the iron lattice, precipitated or separated as an iron carbide, or completely separated as graphite. An illustration of the ranges of temperature and carbon content at which the different situations arise is shown in Figure A.4.

This is a phase diagram. It is a map of the components that are found in near-equilibrium conditions at a point defined by temperature and composition. Carbon stabilises the fcc structure, austenite, at temperatures above 723°C. The solubility of carbon in austenite is significantly greater than in ferrite, so that below this temperature, if the metal is cooled slowly, the carbon is rejected from the metal lattice and separates as iron carbide, called cementite, as the metal takes on the ferrite (bcc) structure.

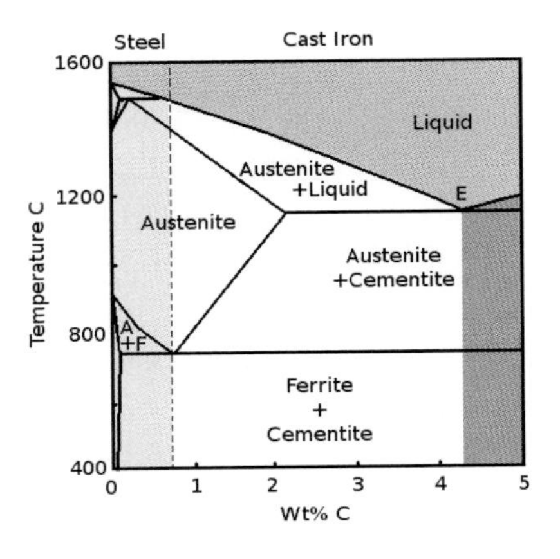

Figure A.4 *The iron carbon phase diagram.*

Steels have carbon contents up to no more than 0.8% – the light grey range in Figure A.4. The top of this range is only found in one important material, patented steel wire, which is processed to give a very strong fibrous structure that is used in wire ropes and was developed in Birmingham. It has an illustrious history.[10] Steels containing more than a small amount of carbon solidify first as ferrite, transform to austenite at a temperature above 1394°C and back to ferrite at a temperature between 910°C and 723°C. The latter is called the eutectoid temperature.

The constituents shown in Figure A.4 are found if the metal is cooled quite slowly from the liquid state. If cooling was infinitely slow, particularly for compositions above the steel range, the carbon would be in the form of graphite. However, in practical circumstances cooling is not very slow and the real interest in useful materials is what structures can be produced by quenching and tempering.

To make the steel hard it is heated to a temperature that ensures that all the material is austenite and then quenched rapidly into water.

Below 723°C the metal prefers to be ferrite, but when there is insufficient time and thermal activation for the carbon to separate out as cementite, it is trapped in a structure that is highly strained and fragmented. However, the metal does not remain as austenite. Below a temperature, referred to as M_s, it does transform to martensite, a distorted form of ferrite, making it hard, but also brittle. Most importantly in so doing it produces a greatly refined structure.

Reheating to a temperature substantially below 723°C allows the carbon to diffuse and form small clusters of cementite, relaxing the metal lattice, but not allowing the structure to coarsen. This results in some loss of hardness but a greater resistance to fracture, because the fragmented structure evolves into a fine grain size when the carbon is precipitated.

If the same basic steel with 0.5% carbon now has some nickel added, it is found that the properties are significantly enhanced. The influence of nickel on iron without the carbon present helps to give some idea of how this occurs. The phase diagram for the binary iron-nickel system is shown in Figure A.5.

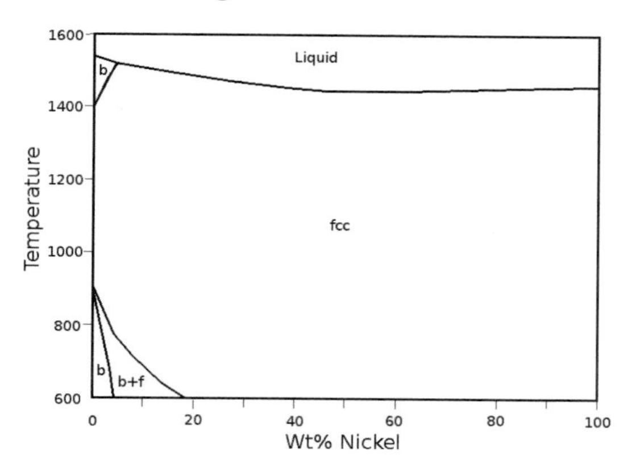

Figure A.5 *The iron-nickel phase diagram, b represents the bcc phase (ferrite), f the fcc phase (austenite).*

The most obvious effect of the nickel is to stabilise the fcc phase, which has the same crystal structure as pure nickel. This results in the temperature at which carbon-containing iron transforms to the bcc form being progressively lowered by increasing amounts of nickel. It is not easy to show the phase structure of three component systems, but it can be taken that the lower of the two horizontal straight lines in Figure A.4 moves to slightly lower temperatures when nickel is added to an iron carbon composition. However, it is also observed that adding nickel to an iron carbon alloy slows down the diffusion of carbon. As noted above, the transformation to ferrite requires the carbon to separate as iron carbide.

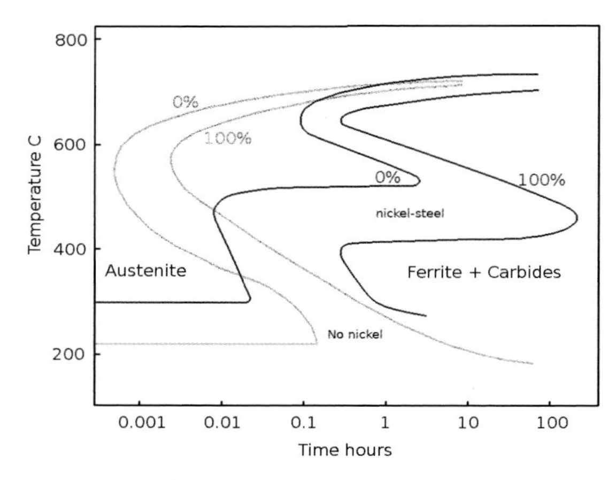

Figure A.6 *Time and temperature of transformation diagrams for a simple carbon steel (in light grey) and one containing nickel (darker shade).*

Transformation to ferrite then occurs at a lower temperature or takes longer to occur. The effect can be shown in a TTT diagram, where the Ts represent time and temperature of transformation. Diagrams for a plain carbon composition and one containing nickel are shown in Figure A.6.

To understand the diagram, consider a piece of the appropriate metal having been heated to 800°C and being allowed to cool. Its temperature after a given time depends on the rate of cooling. Any cooling event can be tracked across the diagram. For the plain carbon steel represented, the first line crossed shows the temperature at which transformation to ferrite and carbide begins. The second line indicates that all of the austenite has transformed. The lower the temperature at which these lines are crossed, the finer the structure will be. If cooling is sufficiently fast that the start line is not crossed until the horizontal line, which is the M_s temperature, is reached, just above 200°C. There is now insufficient thermal energy for ferrite to form by the ejection of carbon from solution in the austenite. That is a thermally activated process, involving carbon diffusion. The carbon is now trapped in the lattice, but the iron needs to take up its low temperature form and tries to do so by undergoing a diffusionless transformation to a distorted form of ferrite, which is martensite. Cooling further below the horizontal line increases the proportion of the austenite that has transformed to martensite.

The diffusionless transformation is achieved by the lattice shearing, that is deforming locally, but in order to produce no total shape change, the original austenite grains generate several different variants of the sheared product so that there is no overall deformation. (Other than the volume change that occurs because the bcc structure has a lower packing density than the fcc). There is, however, a great deal of strain energy trapped in the metal, which can in some cases result in cracks forming.

Each sheared volume will form a ferrite grain when the metal is reheated to a low temperature, allowing the carbon to move sufficiently to form small particles of cementite. The result is a fine-grained ferrite, which is good for toughness, with a fine dispersion of small particles, which provide further strengthening.

It is evident from Figure A.6 that, in the iron-carbon composition, the cooling rate must be very fast to achieve the desired effect, and this is only possible to achieve through the thickness of the metal if it is quite thin. Large components cannot benefit from the improvement that is possible. It is mainly for this reason that nickel is added to the steel used in structural applications. As can be seen, the nickel containing alloy takes much longer to start to transform, which enables the formation of martensite at much slower cooling rates, so that through hardening can be achieved in heavy sections. Nickel increases the hardenability of the material.

The effect of nickel in slowing down the diffusion of carbon and stabilising the austenite, accounts for the double curvature in the nickel steel case, seen in Figure A.6. For the composition illustrated, if cooling to the lower horizontal is accomplished in 10 minutes, the steel transforms to martensite. Cooling more slowly produces a compromise structure, called bainite, in which transformation occurs partly by carbon diffusion and partly by shear. In general, it is preferable to produce martensite.

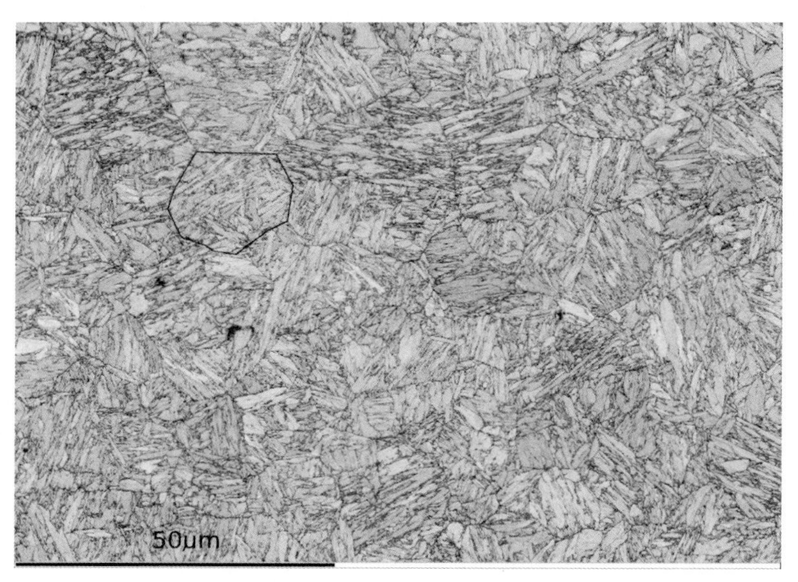

Figure A.7 *0.31%°C steel, water-quenched from 870°C. (Courtesy Professor WB Hutchinson)*

The microstructure in Figure A.7 gives some idea of how quenching to martensite produces, after tempering, a fine-grained ferrite. The martensite comprises many differently oriented laths, formed within the volume of one original austenite grain, (the area of one is outlined in black) of the distorted form of ferrite that has formed by diffusionless transformation.

On reheating the steel to a temperature at which the stable phase is ferrite, the martensite structure relaxes and the individual laths form stress free ferrite grains.

Nickel has a similar effect in cast irons as in steels, although the structures are quite different. High levels of silicon result in carbon being in the form of graphite. A typical carbon content of 3.6% by weight would have about 15% of graphite. With low silicon content, the carbon combines with iron as cementite and the same basic composition would now have about 50% carbide.

Referring to Figure A.4, a composition with 3.6% carbon cooled from the liquid state would start to solidify at just above 1200°C. The first solid is austenite with the maximum carbon that it can contain at that temperature. As the temperature falls to the first horizontal line, called the eutectic temperature, the carbon content in the austenite separating from the liquid increases to a maximum of about 2%. The liquid has increased in carbon content, since the austenite contains less than the average carbon content. Its composition is, in fact, exactly at the eutectic point, E, when the eutectic temperature is reached. Solidification now proceeds at constant temperature until all the metal is solid. It consists of austenite interspersed with an intimate mixture of austenite and cementite.

Continuing cooling slowly, the solubility of carbon in austenite falls with temperature, and secondary cementite particles are precipitated within the austenite, which eventually transforms to ferrite and cementite, the austenite composition having at about 720°C reached the eutectoid composition. (The lower horizontal line is called the eutectoid temperature).

Cooling more rapidly from a temperature around 800°C will result in the formation of martensite, which can be tempered to make the ferrite as fine grained as possible. The entire structure then consists of hard carbides joined by a strong, tough ferrite. The microstructure of a typical material is shown in Figure A.8. It is a native form of a cemented carbide.

Figure A.8 *A white cast iron showing cementite surrounded by ferrite hardened by secondary cementite precipitation.*[11]

(The first ferrous alloy, in which tungsten was introduced into the melt, was a harder form of a similar structure, in which the hard phase is tungsten carbide. Later developments have led to an artificial cemented carbide, in which tungsten carbide particles are mixed with a binder, which is a cobalt containing alloy. Such materials have greater hardness than carbon steels, but are more expensive, and quite unsuitable for large components such as those shown in Figure 7.8).

Nickel acts in the same way in improving the hardenability of the cast irons as in steels. It enables castings to be hardened through much greater section sizes than in the simple carbon compositions.

The effects described above are for white irons. More detail, as for most matters related to nickel, can be found on the Nickel Institute website.[12]

In grey irons, the effects of nickel are more subtle.[13] In the alloys, which contain silicon, the carbon precipitates from the austenite in the form of graphite. Nickel reinforces this effect, although less strongly than silicon, but it does help to avoid the phenomenon of chilling. As noted previously, graphite would in any case be the form carbon takes if the material cooled almost infinitely slowly, so it is the case that even with silicon present, at more rapid cooling rates some of the material can contain cementite. This is most likely to occur at the outer surface, or in thin sections, of a component, so that the grey iron may have a white iron layer, which is harder and more brittle. This is obviously not what the grey iron requires. By reinforcing the effect of silicon, nickel assists in avoiding this problem.

Nickel does, as in most cases, contribute to improved corrosion resistance and to greater strength. In simple grey irons the greater strength is only useful in compression; ductile irons can also benefit in tension.

2.7 The structure of high temperature alloys

Although it had long been known that there was a precipitate formed in the hardening of the strong alloys, like K-Monel, its identity was not known until in 1937 Alexander and Vaughan, working at Birmingham University, produced the first aluminium-nickel phase diagram.[14] It took another twenty years before the way in which this very special phase performed its magic was understood.

The essential character of the nickel-chrome base of the high temperature alloys was discussed in Chapter 9. The formation of an adherent oxide, which is regenerated if broken, on a substrate that is itself resistant to oxidation, is the key to its survival at high temperatures. The base alloys, however, would have inadequate strength were there not a special hardening constituent that is specific to alloys containing nickel. This is usually referred to as gamma prime, written γ'. The name gamma (γ) is usually assigned to the fcc phase, (austenite in steels), so γ' signifies that the hardening constituent is a modification of the γ.

Figure A.9 *Illustrating the interface between the nickel rich matrix and the γ' hardening phase.*

The bead model in Figure A.9 illustrates the nature of the modification. On the right of the model, the pale beads represent a matrix that is predominantly nickel. On the left, the nickel atoms are interspersed with atoms of either aluminium or titanium, shown as grey. These atoms are placed alternately with the matrix atoms along the close packed directions. The two structures fit together perfectly. The second phase is said to be coherent with the matrix. However, if the matrix is deformed, imagine the top layer moving to the right over the middle layer; the periodicity of the hardening phase would be disrupted, effectively altering the crystal structure locally. This requires a high stress. To restore the periodicity of the hardening phase, the atom layer would have to be moved by two atom spacings at a time. Since the deformation of metals occurs by waves (called dislocations) effecting one atom shift per wave and the waves repel each other, the γ' has a very significant hardening effect.

It also has another important attribute. The excellent fit with the matrix, its coherence, ensures that the interface between the matrix and the γ' is of very low energy. In most particle hardened systems, the second phase is not coherent. Iron carbide is a good example. If a steel hardened with iron carbide is heated for a long time at a high temperature, the particles of carbide will grow. The mechanism of growth is by diffusion of the carbon from one particle to a larger one. This reduces the overall surface energy. For a sphere, the total surface area per unit volume is inversely proportional to its radius, so that to reduce the total surface energy of the second phase it will grow progressively if there is sufficient thermal activation. Because they have very low surface energy, and since they are composed of elements of low diffusivity, unlike carbon in iron, the γ' particles are very resistant to growth.

Other aspects of the material processing required to extract the last few degrees of temperature capability out of the nickel-based alloys are not specific to these materials. It has already been noted that coarse grained materials are stronger at high temperatures than those with fine grains. This is because the individual grains are stronger than the interfaces between them and at high temperatures, when there is a lot of atom movement because of the thermal activation, they can slide past each other. Some resistance to grain boundary sliding can be provided by including elements that form precipitates in the grain boundaries, but elimination of the boundaries is the best solution.

As shown in Figure 13.8, the first option is to align the grain boundaries along the length of the airfoil. In this orientation, the boundaries feel the lowest stress possible and the grains are thus least inclined to slide past each other. It is not a perfect solution. The best that can be done is to eliminate the matrix grain boundaries completely.

This is done by making available a single nucleus for the transformation from liquid to solid and then cooling the liquid metal by withdrawing it very slowly from the casting furnace, so that it cools directionally, allowing the seed orientation to propagate through the entire component.

2.8 Carbon-free martensite

Since maraging steels consist of only large atoms that compete for position on the lattice sites defined by the parent metal, iron, the interdiffusion is slow. They solidify in the austenite structure and will retain that structure when cooled, unless cooling is very slow. The stable matrix structure, if cooling were slow enough would be body-centred cubic ferrite, and given time and a little temperature, this is what the structure will seek to become, but the nickel prefers to be in the fcc form and holds up the transformation.

What then is happening on ageing? Clearly the hardening phase, Ni_3Ti is being formed, given time by holding at 480°C for anything between 3 and 12 hours. The nickel had previously been distributed among the other solute atoms and had helped to prevent the matrix from transforming to the bcc structure. Withdrawing nickel from solution resulted in the ferrite stabilising elements predominating and the matrix, which had been a metastable austenite, can now become ferrite. However, there is insufficient thermal activation for this to occur by diffusion, which would involve a lot of large atoms rearranging over significant distances, so what does happen is that martensite forms. The structure becomes a finely divided array of needles of martensite interspersed with platelets of Ni_3Ti that had initially formed, as in the Nimonic structures, in an fcc lattice.

The Ni_3Ti phase is not required to have the additional properties that make it so beneficial in resisting creep. In application at ambient temperature there will be no tendency for the particles to coarsen. The greatest contribution to strength in these materials comes from the ultra-fine micro-structures. These are also good, as was shown in Chapter 7, for toughness. The ageing procedure is a means of creating a fine martensite without the carbon, which is not, in general, helpful in achieving high toughness levels.

Notes and References

Foreword

1. A. Street and W.O. Alexander, 'Metals in the Service of Man', 10th Edition, Penguin Books, London, (1990). ISBN: 10-987-5432-1.
2. J. Mokyr, 'The Gifts of Athena', Princeton Univ. Press, (2002), eISBN: 978-1-4008-2943-9.

Chapter 2

1. Peter Jones, 'Birmingham and the West Midlands in the Eighteenth and Early Nineteenth Centuries', in 'Matthew Boulton – a Revolutionary Player' ed M. Dick, Brewin Books, (2009), p13. ISBN: 978-1-85858-441-6.
2. William F. McDonough, 'The Composition of the Earth', in 'Earthquake Thermodynamics and Phase Transformations in the Earth's Interior' ed Roman Teisseyre and Eugeniusz Majewski, Elsevier, (2000), ISBN: 978-0-12-685185-4. http://quake.mit.edu/hilstgroup/CoreMantle/EarthCompo.pdf
3. John Emsley, 'Natures Building Blocks', Oxford University Press, (2001), ISBN: 978-0-19-850340-8.
4. F.B. Howard White, 'Nickel – a historical review', Methuen, London, (1963).
5. S. Barker, 'Nickel Silver Manufacture' p671 in 'The Resources, Products and Industrial History of Birmingham and the Midland

Hardware District' ed S. Timmins, (1866) Robert Hardwicke, London.

6. Anon, 'The Electroplate Trade and Charles Askin', Birmingham Daily Mail, December 12th, (1878).
7. Edward Benson's son became Archbishop of Canterbury. He had an interesting family that is recorded in Rodney Bolt's book: 'As Good as God, as Clever as the Devil: The Impossible Life of Mary Benson', Atlantic Books, London, (2011). ISBN 978-18435-486-14.
8. J. Percy, 'Metallurgy', John Murray, London, vol 2, (1861). Percy notes that the copper producer H.H. Vivian extracted nickel and cobalt from copper ores using arsenic.
9. 'A Hundred Years of Progress', Henry Wiggin and Company Limited, (1935).
10. Anon, 'The Mond Nickel Company', Company publication (1918).
11. The 1867 Post Office directory of Birmingham lists Evans and Wiggin as metal refiners at Wiggin Street, as well as at the sites that Evans and Askin had started from at 55 George Parade and a later address at Icknield Square. Henry Wiggin had been Mayor of Birmingham in 1864 and had been instrumental in the acquisition of the Wiggin Street site, but Brooke Evans' brother, Alfred, did not die until 1870.
12. Barry A Turner, 'Smethwick Works'. Inco Nickel News.
13. W.C. Roberts-Austen, Proc. Inst. Civil Engineers, (1898), 135, 29.
14. R. Shill, 'Birmingham's Industrial Heritage, 1900–2000', The History Press, (2013), ISBN: 978-0-7509-5411-2.
15. W Markuson and D Baksa, Int. J. History of Engineering and Technology, (2013), 83, (2), 274.
16. W. Weir, 'History of the Weir Group', Profile Books, London, (2008). ISBN: 978-86197-886-8.

    In his history of the Weir Group, William Weir recounts that it lost contact with its agent, a Swiss person, M. Foianesi, during the war with Germany. On re-establishing contact after the war, M. Foianesi visited the company at its head office in Cathcart. During the visit, he asked Kenneth Weir how he should dispose of the stock of nickel that he held in France. This was a surprising question, because shortage of nickel was one of the problems that restricted Germany's war effort in several areas of activity: armour plate, aviation and chemical engineering. It had been assumed that it would have been confiscated. In fact, it turned out to be one of the great unsung manifestations of French resistance. A friend of Foianesi was an official of the French railways. He had the stocks of nickel loaded into railway carriages, which were moved round from time to time from one country siding to another, keeping them out of the hands of the Germans.

17. Before Mond Nickel bought the site, it had been used by the Lion Toffee company. This was where Bluebird toffee was first made, before the business relocated to Romsley, near Halesowen in 1927. There was also an iron foundry shown on the 1887 map, but the business occupying the site before Mond Nickel bought it was the Metallic Seamless Tube Company. That was taken over by Tube investments in 1926.
18. http://www.zen16842.zen.co.uk/WigginSt.

Chapter 3

1. D. Mendeleev, Zeitschrift für Chemie, (1869), 12, 405.
2. J. Emsley, 'Nature's Building Blocks', OUP, Oxford, (2003), ISBN: 0-19-850340-7.
3. L. Pauling, 'The Nature of the Chemical Bond', Phys. Rev. (1938), 54, 899.
4. R.M. Bozorth, Rev. Mod. Phys. (1953), 25, 42.

Chapter 4

1. http://www.silvercollection.it
2. http://www.blackdykebandheritage.co.uk/history/band
3. http://manchesterhistory.net/manchester/gone/higham.html.
4. https://www.youtube.com/watch?v=YpUy1Ra_WN0

Chapter 5

1. Anon, 'The Mond Nickel Company', Company publication (1918).
2. L. Pauling, Phys. Rev., (1938), 54, 899.
3. N.B. Pilling and R.E. Bedworth, J.Inst.Met., 29, (1923), p529.

Chapter 6

1. P.H. Tuck, US Geological Survey, Mineral Commodity Summaries, January (2012), p108.
2. A. Volta, Phil Trans Roy Soc, (1800), 90, 106
3. W. Nicholson and A. Carlisle, Journal of Natural Philosophy, July 1800, iv, 179.
4. H. Bessemer, 'An Autobiography'. Engineering. (1905).
5. G. Bird, Phil,Trans., (1837), 127, 37.
6. For a detailed discussion of the many participants in the development of nickel electroplating see J.K. Dennis and T.E. Such, 'Nickel and Chromium Plating', Butterworth, London, (1982), ISBN: 0-408-01124-6.
7. F.B. Howard White, 'Nickel – a historical review', Methuen, London, (1963).
8. Peck's Circular Trades Directory, 1896-7, W.R. Peck, Birmingham.
9. The current owner of the Canning business claims that the company can be traced back to 1795, no doubt believing the advertisement. It is not credible. W. Canning first appears as the co-owner of a grocery at 1 Kenion Street in the trade directory of 1839. In 1832 the grocery was under the name of Edward Green and in 1835 the grocer was Edmund Gunn, Canning appearing as his partner in the 1839 directory. 1 Kenion Street (later Kenyon Street) is next door to 137 Great Hampton Street, and by 1914 the Canning operation occupied 1 and 2 Kenyon Street and 133-137 Great Hampton Street.
10. N.V. Hybinette, US Patents 579,111, (1897); 805,555, (1905); 805,969, (1905).
11. R.J. Kendrick, Trans. Inst. Metal Finishing, (1964), 42, 235.

Chapter 7

1. K.C. Barraclough, 'Steelmaking Before Bessemer' Vol.2, 125. The Metals Society, London, 1984.
2. J. Stodart and M. Faraday, 'On the Alloys of Steel', Phil Trans Roy Soc. (1822), 112, 253.

3. J. Percy, 'Metallurgy, Iron and Steel', John Murray, London, (1864), 171.
4. J. Riley, J. Iron and Steel Inst., (1889), (No. 1), 45-55.
5. F.B. Howard White, 'Nickel – a historical review', Methuen, London, (1963).
6. 'The Mond Nickel Company Limited', (1918).
7. R.A. Hadfield, 'The History and Progress of Metallurgical Science', https://archive.org/details/historyprogresso00hadf.
8. N.J. Petch, Journal of the Iron and Steel Institute (1953), 173, 25.
9. T. Turner, 'The Metallurgy of Iron', 3rd Edition, Charles Griffin & Co., London, (1908).
10. The effect of magnesium was first published in 1949 in a patent granted to International Nickel, based on work at its laboratory in Bayonne, New Jersey, carried out by a team including a future Chairman of the Company, Albert Gagnebin (1909–1999) K. Mills, L. Aunkst, A. Gagnebin and N. Pilling US Patent No.2, 485,760, (1949).

    The magnesium was introduced using a nickel magnesium alloy. The effect had been anticipated to some extent in the Meehanite process which used calcium silicide and by Henton Morrogh (1917–2003), Director of the British Cast Iron Research Association in Alvechurch, who used cerium. (H. Morrogh and W.J. Williams Journal of the Iron and Steel Institute, 1949, 158, 306).
11. G.W. Form and J.F. Wallace, 'Nickel and Grey Iron Influence on Structure and Properties', Nickel Institute publication 310.
12. Anon, 'Ni Hard Material Data and Applications.' Nickel Institute publication 11017.

Chapter 8

1. C.E. Guillaume, Nobel Lecture, December 11, 1920. https://www.nobelprize.org/nobel_prizes/physics/laureates/1920/guillaume-lecture.pdf
2. J. Hopkinson, Phil. Trans. Roy. Soc, (1889), 180, 443.
3. E.A. Owen, E.L. Yates and A.H. Sully, Proc. Phys. Soc. (1937), 49, 315.
4. P. Chevenard, Comptes Rendu, (1914), 159, 53 and 175. (1917), 164, 916. (1917) 164, 105 and (1917), 165, 59.

5. D. Sobel, 'Longitude', Walker & Co., London, (1995). ISBN: 0-00-721422-7.

Chapter 9

1. Anon, 'Albert Marsh, Inventor, Scientist', Pana News – Palladium, June 5, 1997.
2. R.F. Decker, Journal of Metals, (2006), 58, (9), 36.
3. Anon, 'A Century of Progress'. Henry Wiggin and Co. (1935).
4. https://britainfromabove.org.uk/en/image/EPW054228
5. Lodge's involvement in the history of radio makes his interest in the paranormal more reasonable at that time than it may now seem.
6. https://www.cryptomuseum.com/crypto/colossus/index.htm
7. W. Rosen, 'The Most Powerful Idea in the World', Pimlico, London, (2011). ISBN: 1-10-12024-24.
8. L. Darken and R. Gurry, 'Physical Chemistry of Metals' McGraw-Hill, New York, 1953, page 53.

Chapter 10

1. L.E. Shoemaker and G.D. Smith, 'A Century of Monel Metal', Journal of Metals, (2006), 58, (9), 22.
2. http://www.worldstainless.org/Files/issf/non-image-files/PDF/ISSF_Stainless_Steel_in_Figures_2016_English_Public.pdf
3. 'The Letters of Faraday and Schönbein', ed. G.W.A. Kahlbaum and F.V. Darbishire, Williams & Norgate, London, 1899.
4. W. Borchers and P. Monnartz, German Patent 246,015, (1910).
5. P. Monnartz, Metallurgie, (1911), 8, (7), 161.
6. D. Dulieu, 'Stay Bright', Outokumpo Stainless Ltd, UK (2013).
7. H. Seeböhm, 'On the manufacture of crucible cast steel', Journal of the Iron and Steel Institute, (1884), 2, 372.
8. J. Riley, J. Iron and Steel Inst., 1889 (1), 45.
9. G. Oakes and K.C. Barraclough, 'Steels', Chapter 2 in 'The Development of Gas Turbine Materials' ed. G.W. Meetham, Applied Science Publishers, London, 1981, ISBN: 0-85334-952-5.

Chapter 11

1. H.L. Eiselstein and E.N. Skinner, ASTM STP No. 165, (1954), 162.
2. D.J. Tillack and J.E. Guthrie, Nickel Institute Publication 10071.
3. J.W. Oldfield and B. Todd, British Corrosion Jnl., 26, (3), 173.
4. J.E. Truman and H.W. Kirkby, Metallurgia, Aug, (1965), 67, 273.
5. J. Heselmans and P. Vermeij, Paper 2331, NACE Conference 2013.
6. J. Heselmans private communication to M. White.
7. 'Heat and corrosion resistant castings', Nickel Institute Publication, 266.

Chapter 12

1. M. Faraday, 'Experimental Researches in Electricity, (1849) Google ebook #14986.
2. H.D. Arnold and G W Elmen, Bell System Technical Journal, (1923), 2, (3), 101.
3. T.D. Yensen, US Patent 1,358,810, (1920).
4. P.P. Cioffi, US Patent 1,708,936A, (1929).
5. R.M. Bozorth and J.F. Dillinger, US Patent 2,002,689, (1935).
6. W.T. Lyman, 'Transformer and Inductor Design Handbook', Dekker, New York, 3rd Ed. (2004).
7. T. Mishima, US Patent 2,027,994, (1936).
8. K. Prange, 'Alnico, the Miracle Metal', Premier Guitar, October 21 2009.

Chapter 13

1. G.W. Meetham, 'The Development of Gas Turbine Materials', Applied Science Publishers, Essex, (1981), Appendix 2. ISBN: 0-85334-952-5.
2. N.B. Pilling and P.D. Merica, UK Patent 38,378, (1931).
3. I.L. Dillamore, 'An Industrial Evolution', Completely Novel, (2014). ISBN: 978-8491-4536-7.
4. N.P. Allen, Biogr. Mems Fell. R. Soc. (1972), 18, 476.
5. L.B. Pfeil, UK Patent 583,162, (1940).
6. W. Betteridge and J. Heslop, 'The Nimonic Alloys', 2nd Ed, (1974), Edward Arnold, London, ISBN: 0-7131-3316-3.
7. C.H. White, Chapter 4, page 89 in Meetham, ref 1 above.

8. http://www.engineeringtoolbox.com/fuels-higher-calorific-values-d_169.html.
9. C.G. Bieber and W.F. Sumpter, US Patent 2,570,194, 1951.

Chapter 14

1. P.D. Merica, US Patent 1,572,744 (1923).
2. W. Fragette and J. Mihalisin, ASTM Special Tech. Pub. No.339 (1963), 69.
3. C.G. Bieber, Met Progress, (1960), 78, 99.
4. W. Lowther, 'Iraq and the Supergun', Macmillan, (London), 1991, ISBN: 0-330-32119.

Chapter 15

1. I. Matsamoto, T. Iwaki and N. Yanagihara, US Patent 4251603, (1980).
2. J. Babjak, V.A. Ettel and V. Paserin, US Patent 4957543 A, (1989).
3. G. Sandrock, http://www.energymaterials.co.uk/01_Sandrock.pdf
4. L. Schlapbach and A. Zűttel, Nature, (2001), 414, 353.5.
5. K.J. Kim, K.T. Feldman jnr., G. Lloyd and A. Razan, Applied Thermal Engineering, (1997), 17, (6), 551.
6. B.H. Kang and A. Yabe, Applied Thermal Engineering, (1996), 16, (8), 677.

Chapter 16

1. https://en.wikipedia.org/wiki/Nickel_allergy.
2. http://www.mayoclinic.org/diseases-conditions/nickel-allergy/symptoms-causes/dxc-20267456
3. A. Ölander, Zeitschrift für Kristallographie, (1932), 83, (1/2), 145.

Chapter 17

1. J. Emsley, 'Nature's Building Blocks', Oxford University Press, Oxford, (2001). ISBN: 0-19-850340-8.
2. P. Sabatier, http://www.nobelprize.org/nobel_prizes/chemistry/laureates/1912/sabatier-lecture.html.

3. L.M. Raney, US Patent 1,628,190, (1927).
4. S.Z. Tasker, E.A. Standley and T.F. Jamison, Nature, May 15 (2014),509 (7500), 299.

Appendix

1. M. Faraday Proc Roy. Soc. No.8 December 8, (1831).
2. C.E. Guillaume, Nobel Lecture, December 11, (1920). https://www.nobelprize.org/nobel_prizes/physics/laureates/1920/guillaume-lecture.pdf.
3. T. Mishima, US Patent 2,027,994, (1936).
4. A.J. Bradley and A. Taylor, Proc. Roy. Soc. 56, A159, (1937).
5. D.A. Oliver and J.W. Shedden, Nature, July 30th, (1938), No.3587, 209.
6. W.T. Read, 'Dislocations in Crystals', McGraw-Hill, New York, (1953).
7. A.H. Cottrell, 'Dislocations and plastic Flow in Crystals', University Press, Oxford, (1953).
8. A.A. Griffith, Phil. Trans. Roy. Soc. (1921), A221, 163-198.
9. J. Riley, 'Alloys of Iron and Nickel', J. Iron and Steel Inst., (1889) (No. 1), 45.
10. J. Horsfall, 'The Iron Masters of Penn', Roundwood Press, Warwickshire, (1971) (reprint).
11. T. Turner, 'The Metallurgy of Iron', 3rd Edition, Charles Griffin & Co., London, (1908).
12. 'Ni Hard Material Data and Applications.' Nickel Institute publication 11017.
13. G.W. Form and J.F. Wallace, 'Nickel and Grey Iron Influence on Structure and Properties', Nickel Institute publication 310.
14. W.O. Alexander and N.B. Vaughan, J.Inst.Met. (1937), 61, 247.

About the Author

Ian Dillamore was the last Director of Research of the laboratory that started life as the Mond Nickel Laboratory. He had the misfortune of being obliged to write *finis* to the activities of one of the most distinguished organisations contributing to the development of metallurgical science. How this became necessary is detailed in the historical review in Chapter 1. It was not planned at the time of his appointment. He had previously been employed in Academia and, for a period, in research at another of the great metallurgical laboratories: the BISRA Laboratory in Sheffield. Subsequently, he was offered employment in one of INCO's manufacturing arms, later becoming its Managing Director, and eventually leading its separation from INCO, which retrenched to its core mining business.

Name Index

Subject Index